Entrepreneurial and Startup Opportunities in Veterinary and Animal Husbandry

NIPA® GENX ELECTRONIC RESOURCES & SOLUTIONS P. LTD.
New Delhi-110 034

About the Editors

Dr. Sanku Borkataki, is an academician holding post of Associate Professor in the Sher E Kashmir University of Agricultural Sciences & Technology of Jammu. Dr Sanku is a PhD in Veterinary Parasitology and teaching BVSc&AH, PG and PhD students since 2007. Dr Borkataki has coined innovative ideas of use of insect larvae/maggot in therapeutics and has successfully shown the beneficial activity of maggot. She is very actively involved with student activities and working on generation of innovative idea through the students and incubation of them. She has organized various exhibition programme and help out students to show their idea and creativity in the form of exhibits in various Kisan Mela as well as exhibition.

Dr. Pankaj Goswami, PhD in Veterinary Pathology is working as Professor and Head of the Division of Veterinary Pathology at Sher E Kashmir University of Agricultural Sciences & Technology of Jammu. Dr. Goswami has worked on various infectious diseases of animals and has vast experiences of handling health issue of animal as well as diagnostics in AH and Veterinary sector. Being an academician he has contributed teaching pathology to UG, PG and PhD students in veterinary sciences, and conduct research and extension in his field. He is also involved in the conducting the workshop on startup business in animal husbandry and veterinary sector and thus encourages the young professional to develop entrepreneurship skill through training and awareness. Dr Goswami has published research article as well as popular article in various reputed journal and magazine.

Entrepreneurial and Startup Opportunities in Veterinary and Animal Husbandry

Sanku Borkataki
Associate Professor
Sher E Kashmir University of Agricultural Sciences & Technology of Jammu
Jammu, Jammu and Kashmir, India

Pankaj Goswami
Professor and Head
Division of Veterinary Pathology
Sher E Kashmir University of Agricultural Sciences & Technology of Jammu
Jammu, Jammu and Kashmir, India

NIPA® GENX ELECTRONIC RESOURCES & SOLUTIONS P. LTD.
New Delhi-110 034

NIPA, GENX ELECTRONIC RESOURCES & SOLUTIONS P. LTD.

101,103, Vikas Surya Plaza, CU Block
L.S.C.Market, Pitam Pura, New Delhi-110 034
Ph : +91 11 27341616, 27341717, 27341718
E-mail:newindiapublishingagency@gmail.com
www: www.nipabooks.com

For customer assistance, please contact
Phone: + 91-11-27 34 17 17
Fax: + 91-11-27 34 16 16
E-Mail: feedbacks@nipabooks.com

ISBN: 978-93-58870-88-6

Composed and Designed by NIPA.

Preface

The current scenario in India is very lucrative for starting own venture and entrepreneurship, which is getting more encouragement and boost from all sector as well as from Government of India. Numbers of startup has been developed and initiated through Startup India mission and it is progressing to become big globally. Agriculture combining to business become Agribusiness is a very popular terms now a days and have many resources to make profitable startup and entrepreneurship. Small and medium agribusiness enterprises are very much needed to link smallholder producers to national market, meet food demand and create jobs opportunity. Thus the book **Entrepreneurial and Startup Opportunities in Veterinary and Animal Husbandry** deals with principle and procedure and challenges of entrepreneurship in Agriculture practices in particular reference to Veterinary and Animal Sciences. The innovation and idea to start a business unit and its financial direction are discussed with suitable success stories for personal to take up startup in Veterinary and Animal Husbandry sector. The book mainly focuses the agriprenurship opportunity in Jammu and Kashmir and also cited some of the success stories in the agriculture and veterinary sector. The chapters in the book are written in very simple understandable language and are based on their personal experiences in their academic and extension approach in the subjects.

The present publication is meant for the personal having interest to take startup as well to researcher, students of agribusiness, academician and provide impetus to business personal keen on animal husbandry startup. Hope reader will be benefited going through this book and develop interest in entrepreneurship and startup in animal husbandry sector.

Sanku Borkataki
Pankaj Goswami

Contents

1

Role of Agri-Business Entrepreneurship Models, Policies and Challenges

J.P. Sharma[1] and Anil Bhat[2]

[1]*Former Vice Chancellor, Sher-e-Kashmir University of Agricultural Sciences and Technology, Jammu, India*

[2]*Assistant Professor, Division of Agricultural Economics and Agribusiness Management, Sher-e-Kashmir University of Agricultural Sciences and Technology Jammu, India*

Agribusiness is a term used to describe the sector that encompasses all economic activities that are related to farming, i.e., breeding, crop production/farming, farm machinery, distribution, marketing etc. It relates to industries that are engaged in farming or that produce farm inputs. Examples of agribusiness include farm machinery manufacturing, seed supply, and agrichemicals. It also is used to describe businesses that are involved in the marketing and distribution of farm products. These businesses include warehouses, wholesalers, processors, retailers, and more. Any company that participates in the production, marketing, safety, and distribution of food is involved in agribusiness. More precisely it is defined as a sector that supports the growth of the agricultural industry, which is pivotal to economic growth. It also continues to play a crucial role in the growth of developing countries. Agribusinesses can potentially improve agricultural productivity, which is why governments often offer subsidies to agricultural businesses. Agricultural activities also contribute to an improved system of food security and sustainable food production, as well as income for a majority of the poor in developing countries. However, the activities increase the emission of greenhouse gases and contribute to global warming – which is why innovation is important in the sector to address such problems.

Entrepreneur refers to an individual that has an idea and intends to execute on that idea, usually to disrupt the current market with a new product or service. Entrepreneurship in agriculture can also be defined as the formation of novel economic organization for the intention of growth under risk and uncertainty in agriculture (Dollinger, 2003). Contrary, Gray (2002) defines an entrepreneur as an individual who controls a business with the purpose of growing the business along with leadership and managerial skills necessary for achieving those goals. The most prominent example of entrepreneurship is the starting of new businesses. In economics, entrepreneurship connected with land, labour, natural resources and capital can generate a profit. The entrepreneurial vision is defined by discovery and risk-taking and is an indispensable part of a nation's capacity to succeed in an ever-changing and more competitive global marketplace. Many small-scale farmers and extension organizations understand that there is little future for farmers unless they become more entrepreneurial in the way they run their farms. They must increasingly produce for markets and for profits. Becoming more entrepreneurial can be a challenge for small-scale farmers. They will need help from extension workers and other institutions.

Innovation is defined as **the process of making an idea or invention into a good or service that creates value** and/or for which customers will pay. There are many different types of examples of innovation such as in social innovation, incremental innovation and open innovation among others. The word "innovation" is derived from the Latin verb innovare, which means to renew. In essence, the word has retained its meaning up until today. Innovation means to improve or to replace something, for example, a process, a product, or a service. Therefore we can say:

> *"Innovation is a process by which a domain, a product, or a service is renewed and brought up to date by applying new processes, introducing new techniques, or establishing successful ideas to create new value"*

It requires more creativity and more willingness to take risks than the implementation of typical projects. Entrepreneurs are also innovators. The skills and techniques of innovative thinking are not just vital in work, but useful in everyday life as well, helping us to grow and develop in new situations and think about how to adapt to change more easily. Innovation skills refer to the talent of exploiting new ideas for the purpose of gaining social or economic value. Innovation skills are usually a combination of one's ability to think creatively, problem-solving ability, as well as functional and/or technical abilities.

According to Baragheh in 2009 "Innovation is the multi-stage process whereby organizations transform ideas into new/improved products, service or processes, in order to advance, compete and differentiate themselves successfully in their marketplace". Agricultural innovation is the process whereby individuals or organizations bring new or existing products, processes or ways of organization into use for the first time in a specific context in order to increase effectiveness, competitiveness, resilience to shocks or environmental sustainability and thereby contribute.

A 'value chain' in agriculture describes **the range of activities and set of actors that bring agricultural produce from the point of production to point of consumption**, in which value is added at each stage. At one end of the agricultural value chain are the producers – the farmers who grow crops and raise animals. At the other end are the consumers who eat, drink, wear and use the final products. And in the middle are many thousands of men and women, and small and large businesses. Each person and each business performs one small step in the chain, and each adds value along the way – by growing, buying, selling, processing, transporting, storing, checking, packaging and branding. Other people and other businesses have important roles supporting the chain. Banks provide loans; governments establish laws and policies, and agricultural research organizations develop ways for farmers to more successfully participate in value chains.

The concept of Agri-Business is now changing minds and farmers' also are now becoming an entrepreneur. Small-scale farmers all over the world have shown a remarkable ability to adapt. They look for better ways to organize their farms and enhance their income. They try new crops and cultivars, better animals, and alternative technologies to increase productivity, diversify production, and reduce risk – and to increase profits. They have become more market oriented and have learned to take calculated risks to open or create new markets for their products. They need to be innovative, forward-looking, capability of managing their businesses as long-term ventures and ability to identify opportunities and seize them. The farmer having an entrepreneurial spirit try to act differently and make decisions about the value chain to be followed in his farm to increase the return and profits of the farm business.

Increase in Farmer income is a big task and Government of India is very particular to enhance it through different means. Few procedures are much needed to adapt in the form of models to enhance the income of farmers and these are as:

Promoting Innovations: Innovations are need of the hour to grow as entrepreneur or start-up depending upon the solution to the problems. We need to promote innovations through farmers and young entrepreneurs as these needs to participate actively in designing, testing, adapting and introducing innovations to the agriculture.

Increase in Value Addition: Farmer-entrepreneurs need to realize that capturing value requires producing for buyers and final consumers. But just producing and selling is insufficient. This requires greater understanding and knowledge of value chains and their different elements. It requires a plan for participating further down the value chain. Farmers capture more value by selling their regular fresh produce directly to consumers. Another way is to sell high-value products such as organic products. Instead of producing commodities that are undifferentiated, farmers need to produce commodities that are differentiated such as organic fruit, a unique type of commodity or other specialized products for a limited "niche" market. Another way of adding value is to enter into production and marketing contracts. Contracts are increasingly being used between farmers and input suppliers, farmers and processors and between farmers and other buyers for the production and supply of fresh or processed agricultural produce.

Development and Adaptation of New Technologies: Farmers need to be encouraged to develop and adapt new technologies and share them with other farmers. Extension workers can facilitate partnerships among farmers and between farmers and researchers to work together to identify, develop and test new technologies and practices to improve productivity and profitability.

Use of Information Technology: Now-a-days Information Technology is available and accessible everywhere. Therefore, it is a very powerful way for extension workers to educate and inform farmers about new ideas, technologies and other information. Physical distances and the lack of transportation facilities often limit the ability of extension workers to share information with farmers. Mobile phones, tablets, and computer-based systems can be used to overcome these physical barriers. The challenge is how information technology can be harnessed for the benefit of both extension agents and farmers without compromising the importance of human and unique local factors.

Importance of Agri-Business Entrepreneurship

Agri-Business entrepreneurship is important for national economy in following ways (Sah, 2009).

i. It helps in achieving productivity profit by small farmers and amalgamating them into local, national and international markets.

ii. It helps in decreasing food costs and provides high quality diets to the rural and urban poor in the country.

iii. It accelerates growth, diversifies income and develops entrepreneurial opportunities in both rural and urban areas.

Entrepreneurship Development Opportunities in Agriculture

Agribusiness increases the scenario and opportunities for value addition, packaging, branding, retailing, and exports of agricultural commodities through improved technology and agribusiness management. Indian population directly or indirectly depends on agriculture and now agribusiness is taking the lead and raw material for industries comes from agricultural area only. Now, the time is approaching where agribusiness is most likely to control the Indian economic growth. Presently, there are about 25 per cent to 30 per cent of post harvest losses in fruits and vegetables before reaching the final consumer i.e., due to lack of proper post harvest management practices. However, introduction of agribusiness in agriculture has created lot of changes in business professionalism. The scope and opportunity in the agribusiness has also been increased. The possible opportunities of entrepreneurship in agriculture are:

i. Agro produce processing units: New product is not manufactured but the agricultural produce is processed to form end product. rice mills and dal mills are few examples.

ii. Agro produce manufacturing units: In this category, new products are produced depending upon the agricultural produce as the main raw material. sugar factories and bakery are few examples.

iii. Agro-inputs manufacturing units: These units are engaged in the production of goods either for mechanization of agriculture or for increasing agricultural productivity. Fertilizer and pesticide production units, food processing units and agricultural implements are few examples.

iv. Agro service centres: Agricultural engineering workshops and service centre for repairing, maintenance and serving the agricultural implement used in agriculture are the part of agro-service centres.

Bansal (2011) has proposed some entrepreneurial areas in agriculture:

i. Farming: Crop, dairy/poultry/goat, fish, rabbit, vegetables, flowers, ornamental plants, palmrosa, fodder, sericulture, agro-forestry, beekeeping, mushroom.

ii. Product Marketing: Wholesale, retail, commission agent, transport, export, finance, storage, consultancy.

iii. Inputs Marketing: Fertilizer, agricultural chemicals, seeds, machineries, animal feed, poultry hatchery, vet-medicines, landscaping, agriculture credit, custom service, bio-control units, bio-tech units.

iv. Processing: Milk, fruits, vegetables, paddy, sugarcane, cashew, coir, poultry, cattle, tannery, brewery.

v. Facilitative: Research and development, marketing information, quality control, insurance, energy.

Opportunities for Agri-Business Entrepreneurship in India

Agri-business entrepreneurship is the future of India and Government of India is already focusing on it. Farmers and agri start-ups willing to grow are having number of opportunities for agri-business entrepreneurship and some of them are listed below:

i. Input industry is one of the important industry where demand for agricultural inputs have increased.

ii. Agricultural Biotechnology is having scope in the form of seed production, bio-control agents, industrial harnessing of microbes for different products.

iii. Processing is another area where agri-preneurs are having opportunities and can excel.

iv. Animal husbandry as an industry can become a huge source of generating employment and livelihood. Introduction of new product (Designer Egg, Crossbred cows, Hybrid fowls, etc.) are some of the areas where one can excel as agripreneur.

v. Value addition to the livestock products such as milk, egg, meat, and fish have huge profit potential. Value of the products get increased many folds during processing, and thereby provide excellent returns.

vi. Non-Timber Forest Products are unexploited and therefore is having scope for developing an enterprise.

vii. Beekeeping, apiary and mushroom production can be taken up on great extent in India.

viii. Organic farming has huge potential in India as the pesticide and inorganic fertilizer application are less in India compared to industrial nations of the world.

ix. Huge opportunities for production and promotion of bio-pesticides and bio-control agents for protection of crops.

x. Micro-irrigation systems and labour saving farm equipments have potential in future due to declining groundwater level and labour scarcity for various agricultural operations.

xi. Green House production of vegetables and flowers can play a major role in export of these commodities.

xii. Employment opportunities have increased in marketing, transport, cold storage and warehousing facilities, credit, insurance and logistic support services because of enhanced agricultural production.

Successful Entrepreneurs of Jammu and Agri-Start-Ups Incubated at Agri-Business Incubator, Skuast-Jammu

Various success stories of agri-entrepreneurs have been reported from different sectors and states suggesting their growing importance in agricultural extension (Srinivas IAS et al., 2014). Below are some handpicked success stories about the successful entrepreneurs, farm innovators and agri start-ups incubated at SKUAST-Jammu.

Successful entrepreneurs/ Farm innovators

Mr. Kulbushan Khajuria (58) is a resident of Sohanjana village, Tehsil Mandal, District Jammu of Union Territory of Jammu and Kashmir. He owns 60 healthy milch animals and an automated milk collection system, a well-maintained cattle shed and a feeding area. Currently, his dairy named Krishna Dairy Farm has 48 Holstein Friesians (HF) Cows, nine Jersey cows, two Gir cows, one Shahiwal cow and one Murrah Buffalo. At present, 50 percent of his dairy animals are in the lactating phase and producing 4.5 quintals of milk every day. For marketing of milk, he has established 3 sale counters in Jammu city at Talab Tillo, Shakti Nagar and Channi Himmit Jammu. He is generating a turnover of more than 60 lakhs per year from sale of milk, vermin compost and agricultural products with a net profit of Rs 20-25 lakhs per year. He is providing employment to 7 unemployed youths round the year and 10 persons are engaged for need based services as and when required for cutting of fodder. He is also producing and selling 400-500 qtl. of vermin compost every year through sale counters. To minimize cost of milk production, he is preparing silage on his own farm every year. His visionary and leadership qualities have made him a successful entrepreneur. The success of Mr. Khajuria can be gauged from repayment of loan of 30 lakhs in 5 years that he had taken from the bank at the time of expansion of his dairy farm. He has a clear vision of his venture that helped him to turn the idea of establishing a dairy farm into reality by pooling of his personal resources and availing bank loan.

Sh. Kuldeep Verma S/o Sh. Puran Chand is a resident of village Kotli Charakan, Jammu. He has constructed a pond of 1000 sq. metre. In the initial years, the fish production was less ranging between 200-300 kg/year which was encouraging but not upto his satisfaction. Then, he contacted scientists of KVK SKUAST Jammu for technical details and scientific fish farming techniques. KVK Jammu laid front line demonstration of fisheries at his field and briefed him with all the latest technologies and calendar of activities to follow. He followed all the instructions and achieved fish production of 400 kg during the year 2020. He is earning a net profit of Rs. 59000 per year from sale of fishes. Other than this, he is rearing backyard poultry birds near the fish pond after getting advised from KVK Jammu regarding integrated farming system and excreta of birds will work as feed for fishes and it will reduce his production cost. He is having Sahiwal cow for milk and two dogs for watch during night. His activities are continuously captured and recorded to motivate other youths.

Sh. Dunda Singh a turmeric farmer from Sultanpur village in Kathua district has three roles by profession- a farmer, an entrepreneur and a turmeric Cultivator. For setting up of processing plant he purchased land (1 kanal 3 marlas) @ 1.5 lakhs per kanal with the initial investment of Rs. 7.00 lakhs for creation of infrastructure, 6.92 lakhs on machinery and established the Baba Deep Singh Spices Park with Brand Name BDS at village Kana Chak and registered his products with Food Safety Standard Authority of India (FSSAI) vide Registration No 11017210000050. There is a handsome increase in his income after setting up of the processing plant. By putting all these stepping stones together he started his own turmeric processing plant in 2014 and named it as Baba Deep Singh Spices Park and BDS Haldi is one of the most trusted brands of his processing plant. His family including wife, two sons, two Daughters in laws and maternal uncle plays a major role in all the activities related to the processing of turmeric such washing, boiling, polishing and grinding. He also has seasonal laborers working for him in the processing plant and all the packaging, sealing and grinding work is done by the equal contribution of family members in his factory and home. All these steps taken by him has increased the production of raw turmeric from 60 Quintals to 110 Quintals today.

Startups incubated at Agri-Business Incubator, SKUAST-Jammu

Bhavishya Devgun

Bhavishya Devgun successfully running his start up 'Salutem Ridegear' selling AGRITREAT – An organic agri-input fluid spray offered as an alternative to harmful chemical pesticides & fungicides after conducting successful trials with farmers pan India. He has current sales generation of around 3 Lakhs pan India with distribution partners across 8 states & counting.

Nikky Kumar Jha

Nikky Jha successfully established his start up 'Saptkrishi Scientific Pvt. Ltd. selling SABJI KOTHI – A solar powered, cost effective, IOT enabled, microclimate based wheel mountable Preservator for storage of fresh fruits and vegetables to extend their shelf life by 5-30 days for vendors and small producers. He has exhibited his innovation throughout the country and has been invited by Government of Arunachal Pradesh to run a pilot trial there.

Payal Sharma

Payal Sharma successfully running her start up 'JK Aromatics & Herbs LLP' selling chemical free, natural, sustainable products such as incense, rapid compost, plantable rakhis and chemical free soaps made from flower waste collected from temples/mosques across J&K thereby converting waste to wealth. She has already established a loyal customer base of more than 500 customers.

Vivek Sharma

Vivek Sharma successfully running his start up 'Vatika Agro Services' selling high quality, high yield, disease free seeds & seedlings of vegetables & cash crops developed in soil less media to farmers as well as individual customers including off season seedlings raised in protected structures. He is rapidly establishing a loyal customer base in the Jammu province & is looking to soon expand his operations outside as well.

Kuldeep Dhaliwal

Kuldeep Dhaliwal a progressive farmer successfully running his start up 'Kdhariwal Agrotech & IT Services Pvt. Ltd. Selling CROP RECHARGER – A natural Agri input liquid that enriches the soil with nutrients and increases the nutrient absorption capacity of plants as an alternative to chemical fertilizers. He has won numerous awards & certifications from various govt. & non govt. Organizations for his innovative invention and has a huge customer base of fellow farmers.

Policies

The task of promoting entrepreneurship was earlier given to different departments and government agencies. In 2014, the Prime Minister decided to dedicate an entire ministry i.e., Ministry of Skill Development and Entrepreneurship to build this sector as he felt that skill development required greater push from the government's side. Furthermore, the idea is to reach 500 million people by the year 2022 through gap-funding and skill development initiatives.

India is gradually building a robust startup ecosystem. In order to promote and support entrepreneurs, the government has created a Ministry (department) dedicated to helping new businesses.

Below given is a list of Government schemes/agencies launched to develop and encourage entrepreneurship in India.

- Startup India Seed Fund
- Startup India Initiative
- ASPIRE
- MUDRA Bank
- Ministry of Skill Development and Entrepreneurship
- ATAL Innovation Mission
- eBiz Portal
- Dairy Processing and Infrastructure Development Fund (DIDF)
- Support for International Patent Protection in Electronics & Information Technology (SIP-EIT)
- Multiplier Grants Scheme (MGS)
- Credit Guarantee Fund Trust for Micro and Small Enterprises (CGTMSE)
- Software Technology Park (STP) Scheme
- The Venture Capital Assistance Scheme (VCA)
- Loan For Rooftop Solar Pv Power Projects
- NewGen Innovation and Entrepreneurship Development Centre (NewGen IEDC)
- Single Point Registration Scheme
- Modified Special Incentive Package Scheme (M-SIPS)
- Atmanirbhar Bharat Schemes

The important schemes wherein budding entrepreneurs can get funding or grant-in-aid for establishment and growth of their enterprise are explained below:

RKVY-RAFTAAR

- The Rashtriya Krishi Vikas Yojana (RKVY) is an important scheme of the Government of India, Ministry of Agriculture and Farmers' Welfare (MoA&FW), aimed at strengthening infrastructure in agriculture and allied areas. In order to promote agripreneurship and agribusiness by providing financial support and nurturing the incubation ecosystem, a new component under the revamped scheme RKVY-RAFTAAR has been launched in 2018-19. Sher-e-Kashmir University of Agricultural Sciences and Technology of Jammu got selected as one of the Incubation Centre under RKVY-RAFTAAR project and Agri Business Incubator, SKUAST-Jammu has been established in March 2019.

- It is an initiative to encourage young minds to come with innovative ideas in the field of agribusiness and to develop an agripreneurship ecosystem.
- Under this project there are two Agriprenurship Development Programmes namely **UDGAM** (Pre-Seed Stage Funding) and **PRAGATI** (Seed Stage Funding) with grant-in-aid to the tune of 5 lakhs and 25 lakhs, respectively.

Atal Innovation Mission (AIM)

It is Government of India's flagship initiative to create and promote a culture of innovation and entrepreneurship across the length and breadth of our country. It was setup in 2016. AIM's initiatives have played an important contributory role in the advancement of India from a position of 81 in the Global Innovation Index in 2015 to a position of 48 in 2020. AIM has adopted a holistic approach encompassing schools, universities, research institutions, industry, MSME, NGOs, Ministries, at district, state and national levels.

Major initiatives by Atal Innovation Mission (AIM)

i. Atal Incubation Centres - Fostering world class startups in universities, institutions, private sector adding new dimensions of outcomes based scale up and monitoring to existing incubator model.

ii. Atal New India Challenges - Fostering product and service innovations in the country with national socio economic impact and aligning them to the needs of various sectors/ministries/industry.

iii. Atal Community Innovation Centre - To stimulate community centric innovation and ideas in the unserved/underserved regions of the country including Tier 2 and Tier 3 cities and rural India

iv. Building strategic Innovation partnerships with Public sector, Private Sector, Multinationals, and Global Country to Country Partnerships which include USA, UK, Germany, Russia, Singapore, Sweden, Israel, Denmark, Australia to name a few.

Atal Innovation Mission initiatives are key to further promoting the culture of innovation and giving expression to the innovative and entrepreneurial ideas of startups to enable development of one of the most Innovative Nations of the world.

NIDHI

National Initiative for Developing and Harnessing Innovations (NIDHI) is an umbrella program conceived and developed by the DST for nurturing ideas and innovations (knowledge-based and technology-driven) into successful startups.

NIDHI has following program components

i. Technology Business Incubator (TBI), Seed Support System (SSP), Accelerator, Centre of Excellence (CoE), Promoting and Accelerating Young and Aspiring Innovators and Startups (PRAYAS), Entrepreneur in Residence (EIR),

ii. The NIDHI-EIR programme provides Rs. 30,000 per month as fellowship to an aspiring or budding entrepreneur for pursuing a promising technology business idea.

iii. While Promoting and Accelerating Young and Aspiring technology entrepreneurs (PRAYAS), support upto Rs. 10 lakhs to enable translation of an innovative idea to a prototype is provided.

iv. Inclusive TBI is a three years duration initiative supported by the Department of Science Technology DST for educational institutions who are likely to foster innovation and entrepreneurship culture among the students, faculties, entrepreneurs, and nearby communities.

Startup India Initiative

The Prime Minister of India launched the Startup India Initiative in the year 2016. The idea is to increase wealth and employability by giving wings to entrepreneurial spirits. The government gives tax benefits to startups under this scheme and 798 applicants have made use of this scheme to date. The Department of Industrial Policy and Promotion is maintaining this initiative and is treating it as a long term project. Moreover, the overall age limit for startups has been increased from two years to seven years and for biotechnology firms, the age limit is ten years from the date of incorporation. It is one of the best government-sponsored startup schemes for entrepreneurs as it provides several concessions.

Startup India Seed Fund

On 16 January 2021, Prime Minister Sh. Narendra Modi announced the launch of the 'Startup India Seed Fund' — worth INR 1,000 crores — to help startups and support ideas from aspiring entrepreneurs. Prime Minister said that the government is taking important measures to ensure that startups in India do not face any capital shortage. Financial assistance to startups is also the part of the initiative for proof of concept, prototype development, product trials, market entry, and commercialization. Under the initiative, the Seed Fund is disbursed to eligible startups through eligible incubators across.

Challenges

Agripreneures are facing various challenges in the execution of their plans to develop new start-ups or any other support businesses. The basic problem can be the literacy level of the agripreneurs, farmers or rural people and another strong chance is lack of innovative mind. This acts as root cause for majority of challenges mentioned below. Major challenges faced by the agripreneurs are inadequate infrastructural facilities, lack of innovative entrepreneurial abilities among people, declining interest in agricultural allied services, lack of skilled and managerial manpower, marketing problems, lack of awareness about career in agripreneurship, lack of budget, lack of strategy, innovation initiatives with no innovation strategy, fear of making mistakes, cognitive barriers etc.

Conclusion

Agrientrepreneurship is the need of the hour to make agriculture a more attractive and profitable business enterprise. Agriculture has great scope for entrepreneurship and innovative start-ups and it needs useful management of production and marketing parameters. Agriculture in developing countries is not successful to produce enough income to lift people out of their poverty and scarcity. Thus, innovation and agriculture definitely generates a solution for growing household incomes. The excellent management and entrepreneurial expertise infused with government measures would facilitate success of the growing needs of agri-businesses. Innovation, confidence, risk bearing, honesty and vision are the parameters to be adopted and follow for becoming a successful agrientrepreneur. Agrientrepreneurship provides direct employment especially in rural areas and contributes to the national income. Value addition of products offer entrepreneurs with high profit and return-on-investment which otherwise farmers is not able to receive. Therefore, farmers by becoming entrepreneurs have to take benefit of change in consumer demand and satisfy consumers' needs with value-added products.

References

Bansal A. 2011. Agri-Business in India–Vision 2020. International Journal of Multidisciplinary Research, 1(2):62-72.

Baregheh, A., Rowley, J. and Sambrook, S. 2009. Towards a multidisciplinary definition of innovation. Management Decision, 47(8):1323-1339.

Dollinger, S. J. 2003. Need for uniqueness, need for cognition and creativity. The Journal of Creative Behavior, 37(2): 99–116.

Gray, C. 2002. Entrepreneurship, resistance to change and growth in small firms. Journal of Small Business and Enterprise Development, 9(1):61-72

Kashyap, P., Prusty, A.K., Panwar, A.S., Kumar, S., Punia, P., Ravisankar, N. Kumar, V. 2019. Agri-Entrepreneurship: Challenges and Opportunities, Today and Tomorrow's Printers and Publishers, New Delhi

Sah P, Sujan D K and Kashyap S K. 2009. Role of agripreneurship in the development of rural area, Paper presentation in ICARD at Banaras Hindu University, Varanasi.

Srinivas, IAS., B. Shekara, P. C., Murthy, L. and Sahare J. 2014. Agri-Clinics and Agri-Business Centres Success Stories of Agripreneurs. National Institute of Agricultural Extension Management (An Organization of Ministry of Agriculture, Govt. of India). Rajendranagar, Hyderabad.

2

Emerging Opportunities in Agribusiness

Anil Bhat

Assistant Professor, Division of Agricultural Economics & Agribusiness Management, Sher-e-Kashmir University of Agricultural Sciences and Technology, Jammu, India

Introduction

Everyone is familiar with the term start-up or entrepreneurship which refers to a process by which individuals start and manage their businesses and industrial enterprises. Entrepreneurs are people who take risks and invest their capital in businesses and industrial enterprises. Entrepreneurial in terms of personal development in terms of self-determination create take risks Therefore, anyone who exhibits the characteristics of self-development, creativity, personal decision-making, and risk-taking can be called a person with entrepreneurial characteristics. When these characteristics are exhibited by someone who is running a business, they can be called an entrepreneur. Entrepreneurship is the ability and willingness to develop, organize and manage a business enterprise, with all its uncertainties, in order to make a profit. The most prominent example of entrepreneurship is starting a new business. In economics, business activities involving land, labor, natural resources, and capital can generate profits.

> *"A business vision is defined by discovery and risk taking and is an integral part of a country's ability to succeed in an ever-changing and more competitive global market".*

Agribusiness is actually an innovative concept which unites agriculture with business sector. It is mainly meant to elucidate the problem of wastage of the farmer's products during the years of abundant production. Moreover, production is closely related to the marketing of the product and unless marketing is not proper, production is of no use. Marketing includes transformation, transportation and transfer of all the agricultural commodities to the consumer. In addition, farming is served by a large number of industries such as input, output and processing industry that manufacture and distribute quality goods and other farm supplies used in agriculture. Farmers are supposed to buy agricultural inputs such as fertilizers, seed, petroleum products, farm machinery, chemicals, and other farm supplies and services which are essential or required for the farm production. As far as the link between farmers and food wholesalers and retailers is concerned, food processors take the lead. There are processing firms that add form utility to the raw farm product, for example, the transformation of apple into apple juice or jam. Retailers come into the picture at the bottom of the value chain who sells directly to consumers. A retail food chain is a group of stores owned and operated by the same firm. The major impact of the retail chain store movement was on wholesale operations. Chain stores combined wholesaling with retailing, and to a lesser extent, processing, and were able to lower costs. Such events have significant impacts on human beings and can be understood and predicted better with sound education training in agricultural economics and agribusiness especially when supplemented by a good background in technical agriculture. Agribusiness is a multidisciplinary field of study aimed at providing students with a number of scientific, technical and economic disciplines.

John H. Davis of Harvard University first used the term agribusiness in 1955. In 1980s it was given three connotations: (1) synonymous with term agriculture, (2) synonymous with agricultural economics and (3) a modified concept of agriculture, excluding farming, or the off-farm aspects of agriculture.

The concept of Agri-Business is now changing minds and farmers' also are now becoming an entrepreneur. Small-scale farmers all over the world have shown a remarkable ability to adapt. They look for better ways to organize their farms and enhance their income. They try new crops and cultivars, better animals, and alternative technologies to increase productivity, diversify production, and reduce risk and to increase profits. They have become more market oriented and have learned to take calculated risks to open or create new markets for their products. They need to be innovative, forward-looking, capability of managing their businesses as long-term ventures and ability to identify opportunities and seize them. The farmer having an entrepreneurial spirit tries to act differently and make decisions about the value chain to be followed in his farm to increase the return and profits of the farm business.The challenge of seeking affordable, accessible, easy-to-use technologies, products or services to enhance income and employment in India creates opportunities for young entrepreneurs to establish their startups to provide timely solutions.

Agri-business is the sum total of all operations involved in the manufacture and distribution of farm supplies, production activities on the farm, storage, processing and distribution of farm commodities and items made from them (*John David and Gold Berg)*. Agribusiness transactions may involve either an input or a product or service and include items such as:

1. Productive resources (feed, seed, fertilizer, equipment, energy, pesticides, machinery, insecticides, vermin-compost, organic manure, etc.)

2. Agricultural commodities – (raw and processed agricultural commodities)
3. Facilitative services (credit, insurance, marketing, storage, processing, transportation, packing, packaging, advertising, distribution, consultancy etc.

Agribusiness is sum total of all operations or activities involved in the business of production and marketing of farm supplies and farm products for achieving the targeted objectives. Individual agribusinesses may sell items to farmers for production; provide services to other agricultural businesses; or be engaged in the marketing, transportation, processing, and distribution of agricultural products. Agriservice is activities of value to the user or buyer. The activities are an intangible product. Marketing is providing the products and services that people want when and where they want them.

According to the classic definition agribusiness involves the following branches:

1. Supplying sector of inputs for agriculture and food industry (specialized engineering industries, chemistry, energy, etc.)
2. Agricultural primary production
3. Feed industry
4. Services for agriculture and food industry (supplies, purchasing, maintenance, improving and seed production, breeding services, applied research, education, consulting etc.)
5. Food production and other processing industry
6. Food trade and public catering.

Scope for Agribusiness in India

1. India has diverse climatic conditions for production of various types and varieties of fruits and vegetables such as tropical, subtropical.
2. There is growing demand for agricultural inputs like hybrid seeds, fodder, organic fertilizers etc.
3. Export can be harnessed as a source of economic growth. As a signatory of World Trade Organization, India has vast potential to improve it present position in the World trade of agricultural commodities both raw and processed form. The products line include cereals, pulses, oilseeds and oils, oil meal, spices and condiments, fruits and vegetables, flowers, medicinal plants and essential oils, agricultural advisory services, agricultural tools and implements, meat, milk and milk products, fish and fish products, ornamental fish, forest by products etc (Sundar, 2016)

4. Production of vegetables and flowers under green house conditions can be taken up to harness the export market.
5. Trained human resources in agriculture and allied sciences will take on agricultural extension system due to dwindling resources of state finance and downsizing the present government agricultural extension staff as consulting services.
6. Development in agriculture increases the employment opportunities in the country.
7. Livestock in the country can provide one of the successful enterprises.
8. Bee keeping and apiary is one of the profitable enterprise.
9. As India is rich in forest resources, it provides opportunities in this sector also.

Emerging Areas of Agribusiness

India has various areas where agribusiness can be done successfully. Organic farming, fruit juices, fruit pulps, flavours, candies, pickled products, frozen food items, processed milk products, value added products, organised food retailing, mushrooms, medicinal plants are few of them. They are discussed below in detail:

Indian Organised Retail Market

Organised retailing refers to trading activities undertaken by licensed retailers, that is, those who are registered for sales tax, income tax, etc. These include the corporate-backed hypermarkets and retail chains, and also the privately owned large retail businesses. In other words, it is a network of similarly branded stores with an element of self-service. Organised retail in India today holds only a fraction of the market share potential in India. In 2001-02, organized retail trade in India was worth Rs 11,228.7 billion. It has risen from 0 to 6 per cent in a very short period mainly on volumes and not a value-driven growth (Gupta, 2012). (According to McKinsey & Company report titled 'The Great Indian Bazaar: Organised Retail Comes of Age in India', organised retail in India is expected to increase from 5 per cent of the total market in 2008-09 to 14-18 per cent of the total retail market and reach US$ 450 billion by 2015-16. The food and grocery constitutes the highest retail volume and this share has shown a tremendous growth over the years. According to NSSO 60th round, 54 per cent of the rural and 42 per cent of urban expenditure was on food. The second largest share is commanded by the apparels. Clothing and textile is the largest organised market and is dominated by retailers like Pantaloon, Westside, Globus, Koutons. This owes to the increasing disposable incomes and changing lifestyles. Industry trends for retail sector indicate that organized retailing has major impact in controlling inflation because large organized retailers are able to buy directly from producers at most competitive prices.

Hydroponics: Hydroponics is the cultivation of plants by placing the roots in liquid nutrient solutions rather than in soil (soilless growth of plants and without conventional soil). It uses artificial means to support and nutrient solutions to provide the required nutrition to plants. A person interested in such type of technology can start his business with hydroponic retail store. Moreover this area is gaining momentum for the last few years.

Vermicompost production

Vermicomposting is a simple biotechnological process of composting, in which certain species of earthworms are used to enhance the process of waste conversion and produce a better product. It has become major component of agri-business models with low investment cost.

Hardening nurseries: Hardening of the plants in the nursery includes "Any treatment that makes the tissues firm to withstand unfavourable environment like low temperature, high temperature and hot dry wind". Such type of business is also gaining importance as far as agribusiness is concerned.

Processing Industry

Fruit and vegetable processing, including dehydration, canning, aseptic packaging, packing, processing of underutilized fruits, and processing for other products walnut, grape raisin, air-dried fruits, fruit candies and spices' powders etc. Processing of maize for starch and feed through improved mini/ small mills and dry milling plants, processing of sugarcane for various jaggery products like spiced jaggery, powdered jaggery, and jaggery cubes, processing of dairy products and many more are part of the processing industry. As demand for processed foods is increasing day by day so the opportunity for agribusiness is also increasing because all the processing industries falls under agribusiness ventures.

Milk Cooperatives

Now-a-days selling of mlik on cooperative basis also forms the part of agribusiness model. Dairy farmers are constituting a group for collection of milk at their own. The milk is than passed through chilling process at collection centre and from the centre it is transported to cooperatives for further processing of the milk.

Packing and Packaging

Package attracts consumer's attention to particular brand, enhances its image, and influences consumer's perceptions about product (Rundh, 2005). Also package imparts unique value to products (Underwood et al., 2001; Silayoi and Speece, 2004), works as a tool for differentiation, i.e. helps consumers to choose the product from wide range of similar products, stimulates customers buying behavior (Wells et al., 2007). Thus package performs an important role in agribusiness and could be treated as one of the most important factors influencing consumer's purchase decision. In this context, seeking to maximize the effectiveness of package in a buying place, the researches of package, its elements and their impact on consumer's buying behavior as far as agribusiness is concerned became a relevant issue. A packaging that is well designed for its marketing function helps to sell the product by attracting the attention and expressing the product positively to the ultimate potential consumers. To achieve the communication goals effectively and to optimize the potential of the packaging in a buying place, marketers must recognize the attributes of a packaging and their impact on consumer's purchase decision and integrate these perceptual attributes into their packaging design.

Potential of Rural Markets

The rural markets are growing fatty compared to the urban markets. The potentiality of rural markets is said to be like a 'woken up sleeping giant'. These facts are substantiated in a study of market growth conducted by various researches. In recent years, rural markets have acquired significance in countries like China and India, as the overall growth of the economy has resulted into substantial increase in the purchasing power of the rural communities. The reasons why companies are going rural are manifold. Higher rural income driven by agricultural growth, increasing enrolment in primary schools, high penetration of TV and other mass media have increased the propensity to consume branded and value added products in rural areas. Marketers and manufacturers are increasingly aware of the burgeoning purchasing power, vast size and demand base of the once neglected Indian hinterland. Efforts are now on to understand the buying behaviour of rural consumer and to walk their walk and talk their talking talk. The marketing mix of many companies is now being tailored to rural taste and lifestyle.

Meat Consumption

Food consumption is a basic action contributing to survival of human beings. Therefore, the understanding of the food culture creates a better understanding and more intimate relationship among countries and people. The improvement in agriculture, allowed some fundamental changes in the human way of life. Animal derived food goods are the most resource intensive foods (Chemnitz and Becheva, 2014; Connor and Mínguez, 2012). Animal protein foods are at the top of the food chain in relation to the resources (Chemnitz and Becheva, 2014). In many developing nations, it is expected that global meat consumption will continue to rise (Buttriss, 2011; Connor and Mínguez, 2012). It has been suggested that if meat consumption patterns continue along the same path, then the consumption rates will be 72 per cent higher than 2000 levels in the year 2030. Demand for meat is growing in rapidly dev eloping countries, particularly India, China and Brazil (Cheminitz and Becheva, 2014). Since 1970s, there was a reduction in intake of whole grain cereals, pulses, fruits and vegetables in India while intake of meat products have been increased (Misra, 2011; Popkin, 2002; Popkin, 2012).

Floriculture

Floriculture is now profitable venture worldwide as it is one of the fastest growing crop trends based on its demand. People are using flowers in their day to day activities also, moreover, it is mostly used in marriages and functions as loose flowers. Demand for cut dry flowers has also increased tremendously for so many years.

Export of Fruits and Vegetables

Now-a-days government is providing support to the farmers for exporting their product to get better price for their produce. Farmers' can start their own enterprise/ business for export of good quality fruits and vegetables where demand for the same exists.

Frozen Foods

Due to the paucity of time, demand for frozed food items is increasing continuously. An entrepreneur interesting in such type of business will definitely earn a handsome income.

Mushroom Production

Mushroom production business can be a mean of big profit in just a few weeks with considerable low investment to start a business. Button mushroom is demandable now a days. Oyster mushroom is also important as far as the environment and temperature of Jammu is concerned. Post harvesting of the same is very easy. Oyster mushroom is also used in the pharmaceutical industry for boosting of immune system and control of diabetes. Soap and detergent companies are also using it in their products. We can form powder of oyster mushroom after drying it and same can be consumed with milk.

Juice-Jam-Jelly Processing Industry

Processing of fruits into juice, jam and jelly has the huge business opportunities as these products fall under ready to drink or ready to eat products category. The demand for these products is at its peak in the current time and so the business opportunity is also high.

Besides above agribusiness opportunities discussed in detail, few are given below:

Flour milling, groundnut processing, cashew-nut processing, snail farming, sunflower farming, bee-keeping, retailing, organised retailing, fish farming etc.

Conclusion

From the above discussion and trends discussed, it is clear that agribusiness is not only important but lifeline to sustain the livelihood of farmers in world as well as in India. At present about 7,20,343 agribusiness firms are working in different sectors in the country. The trends in global agribusiness are changing at a rapid pace and the traditional practices in agriculture seem obsolete today. Now this sector is more concerned with delivering customer value and diversification rather than increasing quantity of production only. The consumer is more conscious of a number of product attributes before making a decision about the product like low nutritive value, good taste, convenient packaging, easy availability and "Green" products. India is a developing nation and scope of business opportunities is more, therefore, making way of agribusiness ventures also. Government is supposed to take initiatives to boost this sector and overcome constraints which concerned people are facing in start of business.

References

Chemnitz C., and Becheva S. 2014 Meat Atlas 2014.

Connor D. J., Mínguez M. I. 2012 Evolution not revolution of farming systems will best feed and green the world Global Food Security 1, 106–113.

Buttriss J. L. 2011. Feeding the planet: an unprecedented confluence of pressures anticipated Nutrition Bulletin 36, 235 – 241.

Gupta, U. (2012). Customer Loyalty towards Kiranas in Competitive Environment; A Case study International Journal of Marketing and Technology, II (8), 249-268.

Misra A., Singhal N., Sivakumar B., Bhagat N., Jaiswal A., Khurana L. 2011. Nutrition transition in India: secular trends in dietary intake and their relationship to diet-related non communicable diseases J. Diabetes 3, 278-292.

Popkin B. M. 2002. An overview on the nutrition transition and its health implications: the Bellagio meeting Pub. Health Nutr. 5, 93-103.

Popkin B.M., Adair L.S., Ng S. W. 2012. Global nutrition transition and the pandemic of obesity in developing countries Nutr. Reviews 7, 03-21.

Rundh, B. 2005. The multi-faceted dimension of packaging. British Food Journal, 107 (9): 670-684.

Silayoi, P., and Speece, M. 2004. Packaging and purchase decisions: An exploratory study on the impact of involvement level and time pressure. British Food Journal, 106 (8), 607-628.

Sundar, I. 2016. Agribusiness Scope, Opportunities and Challenges in India. International Journal of Economic and Business Review. 4(7):171-178

Underwood. R. L., Klein, N. M. and Burke, R. R. 2001. Packaging communication: Intentional effects of product imagery. Journal of Product and Brand Management, 10 (7): 403-422

Wells, L. E., Farley, H. and Armstrong. G. A. 2007. The importance of packaging design for own-label food brands. International Journal of Retail and Distribution Management, 35 (9), 677-690.

3

Role of Jammu & Kashmir Entrepreneurship Development Institute (JKEDI) in the Development of Entrepreneurship

Vinod Kumar

Incharge Centre for Business Development, JKEDI, Jammu, India

An **entrepreneur** is an individual who creates a new business, bearing most of the risks and enjoying most of the rewards. The process of setting up a business is known as entrepreneurship. The entrepreneur is commonly seen as an innovator, a source of new ideas, goods, services, and business/or procedures.

Entrepreneurship is an act of being an entrepreneur, or "the owner or manager of a business enterprise who, by risk and initiative, attempts to make profits". Entrepreneurs act as managers and oversee the launch and growth of an enterprise. Entrepreneurship is the process by which either an individual or a team identifies a business opportunity and acquires and deploys the necessary resources required for its exploitation.

An entrepreneur uses their time, energy, and resources into creating value for others. They are rewarded for this effort monetarily and therefore both the consumer of the value created and the entrepreneur are benefitted.

Characteristics of Entrepreneurship

Not all entrepreneurs are successful; there are definite characteristics that make entrepreneurship successful. A few of them are mentioned below:

- **Ability to take a risk-** Starting any new venture involves a considerable amount of failure risk. Therefore, an entrepreneur needs to be courageous and able to evaluate and take risks, which is an essential part of being an entrepreneur.
- **Innovation-** It should be highly innovative to generate new ideas, start a company and earn profits out of it. Change can be the launching of a new product that is new to the market or a process that does the same thing but in a more efficient and economical way.

- **Visionary and Leadership Quality-** To be successful, the entrepreneur should have a clear vision of his new venture. However, to turn the idea into reality, a lot of resources and employees are required. Here, leadership quality is paramount because leaders impart and guide their employees towards the right path of success.
- **Open-Minded-** In a business, every circumstance can be an opportunity and used for the benefit of a company. For example, Paytm recognised the gravity of demonetization and acknowledged the need for online transactions would be more, so it utilised the situation and expanded massively during this time.
- **Flexible-** An entrepreneur should be flexible and open to change according to the situation. To be on the top, a businessperson should be equipped to embrace change in a product and service, as and when needed.
- **Know your Product-**A company owner should know the product offerings and also be aware of the latest trend in the market. It is essential to know if the available product or service meets the demands of the current market, or whether it is time to tweak it a little. Being able to be accountable and then alter as needed is a vital part of entrepreneurship.

Importance of Entrepreneurship

- **Creation of Employment-** Entrepreneurship generates employment. It provides an entry-level job, required for gaining experience and training for unskilled workers.
- **Innovation-** It is the hub of innovation that provides new products, services, ventures, market, technology etc. and increases the standard of living of people.
- **Impact on Society and Community Development-** A society becomes greater if the employment base is large and diversified. It brings about changes in society and promotes facilities like higher expenditure on education, better sanitation, infrastructural development etc. Therefore, entrepreneurship assists the organisation towards a more stable and high quality of community life.
- **Increase Standard of Living-** Entrepreneurship helps to improve the standard of living of a person by increasing the income. The standard of living means, increase in the consumption of various goods and services by a household for a particular period.
- **Supports Research and Development-** New products and services need to be researched and tested before launching in the market. Therefore, an entrepreneur also dispenses finance for research and development with

research institutions and universities. This promotes research, general construction, and development in the economy.

Jammu & Kashmir Entrepreneurship Development Institute (JKEDI)

JKEDI has been established by the Government of Jammu and Kashmir in March 1997 to effectively enable entrepreneurship development in the Union Territory of Jammu & Kashmir. The institute started its regular activities from February 2004 and has positioned itself as a learning centre par excellence with state of the art regional centers across the Union Territory. Besides, JKEDI District Nodal Officers are in all of the districts enabling entrepreneurship and promoting development at the grassroots.

Role of JKEDI in Entrepreneurship Development

For holistic and sustainable entrepreneurship development, there is always a need for JKEDI to step out of its main campuses into the districts, and even deeper into the community. By combining community organizing skills and participatory training practices, JKEDI professionals work at the grassroots to create enterprise awareness and readiness among the youth of Jammu and Kashmir. The Institute has a rich intellectual pool of faculty and experts who work selflessly in the areas of entrepreneurship education, trainings, research business assistance support services & Startup financing etc.

The Institute so far has implemented various central as well as state sponsored schemes which inter alia include Seed Capital Fund Scheme (SCFS) and Youth Startup Loan Scheme (YSLS) under the auspices of Department of Labour & Employment, Govt. of J&K, Term Loan & Educational Loan Schemes of National Minorities Development Finance Corporation (NMDFC) under the auspices of Ministry of Minority Affairs, Govt. of India and Himayat self-employment scheme under the auspices of Ministry of Rural Development, Govt. of India. The Institute so far has trained more than 38,000 youth under different schemes and till date more than 13,500 enterprises have been set up successfully under various schemes across the length and breadth of the Jammu & Kashmir.

Besides, JKEDI has also been designated as the nodal agency for JK Startup policy 2018. The Start-up Policy-2018 rolled out in September 2018 envisions the creation of 500 new Start-ups in J&K in next 10 years, besides establishing Incubators, Fabrication and Innovation labs. The Policy also contemplates to create two Start-up Hubs in Kashmir and Jammu besides setting up of Angel Networks, Entrepreneurship Development Cells, Sector/Actor Collaboration, Strong incentive Structure and other Academic Interventions

Conclusion

The aim of JKEDI is to develop entrepreneurial ecosystem in the Union Territory of Jammu and Kashmir. The Institute maintains and constantly improves its distinctiveness as a premier Multidisciplinary Development Institution through cross-cutting approaches of awareness, training, consultancy and investment in entrepreneurship education and research.

The entrepreneur with his vision and ability to bear risk can transform the economic scene of the country. They play a vital role in initiating and sustaining the process of economic development of a nation. The Institute's activities are thus targeted to stimulate educated unemployed youth for adopting entrepreneurship as a career option and thus resolving the problem of unemployment in the Jammu & Kashmir.

4

Opportunities for Startups in Livestock Sector

Manish K. Chatli[1] and Pratikshya Panda[2]

[1]Director, ICAR-Central Institute for Research on Goats, Makhdoom, U.P.
[2]Assistant Professor, Department of Veterinary and Animal Husbandary Extension, College of Veterinary Science, GADVASU, Rampura Phul

Current scenario of Startups in India

India, currently is being evolved as a hub for Revolution, innovation and Startups as it holds 3rd rank in global startup ecosystem worldwide after US and China. The flagship initiative of Government, Startup India, is intended to construct arobust eco-system for fostering innovations and startups in the country. Through this initiative, the Government aims to empower the growth of startups through new innovations and strategies. As per the reports of Startup India, the startup sector is expected to mark steady annual growth of 12-15% (startupindia.gov.in). By the year 2018, India already had a list of about 50,000 startups. As per the data, more than 800 technology startups are being established every year (NASSCOM, 2015). The annual growth has increased to 15% and the number of female entrepreneurs has also increased to 14% (startupindia.gov.in). As of December 2021, the country is home to 81 unicorns which has a total estimate of $ 274.17 Bn. Out of these unicorns, 44 unicorns came into existence in 2021 itself, which was a record year for the country. These Figures show that startups have exploded in the country in recent years.

Scope for Startups in Indian Agricultural Systems

Agriculture plays a pivotal role in India's economy. More than 58% of the rural households are primarily dependent on it for generating livelihood. Similarly, the Indian grocery and food market is the 6th largest market in world with retail backing 70% of the sales. The food processing industry of India contributes around 8.39% of Gross Value Added (GVA) in Agriculture (nmoop.gov.in). This sector plays a significant role in rural livelihood, employment and national food security. The share of this sector in gross domestic product

(GDP) has touched around 20 % making it the exclusive contributor in GDP performance during 2020-21as per the Economic Survey 2020-2021(Shagun, 2021). Still, this is one of the uncertain sectors because its dependency on various factors like market, weather and topographical conditions. Many efforts are being undertaken to give this sector a boost towards its fullest potential. One of the effective ways to do this is through improvements and innovations in agriculture technology. Which further create a huge scope for Startups in this sector. Transforming agriculture to Agribusiness is one of the important strategies. As per reports, startups have undoubtedly driven growth (Krishnan, 2018) and also involved in revamping the agriculture sector of the country (Mattoo, 2018). Despite the widespread pandemic, agritechstartups reaped $152 million in funding through 30 deals in 2020 (thestartuplab.in). The Food Marketplace/Commerce sector has highest start-up investors i.e. 40%, followed by Ag Biotechnology (22%) (AgtechFunding Report, 2016).

Animal Husbandry as a Vital Subsector of Agriculture

Livestock is a vital component of the Indian agriculture as it plays a substantial role in providing food, power and income generation to the poor. It alone contributes 25% to the total agricultural GDP. It also contributes to 4.11% to the National GDP. Around 16.44 million workers were engaged in the activities of farming of animals, fishing and mixed farming etc. (As per the NSS 68th Round Survey, 2011-12). As per the census reports, the country has 535.8 million of total livestock population and 851.8 million of poultry (20th Livestock Census, 2019). Hence, there is a huge scope for the growth in the sector, provided that the industry becomes more organized.Currently most of the livestock production is managed by poor farmers with limited resources which is not likely to provide direct benefits from advanced technologies. Productivity of livestock is very low which is mainly due to poor nutrition, improper management and infectious diseases whereas the demand for livestock products is expanding exponentially. Thus, in order to meet the demand of the livestock industry needs to shift to a larger and effective production system.

Scope for Startups in Livestock Sector

Livestock sector in India is suffering from a varied problem like unskilled workers, use of outdated equipment, improper infrastructure, and lack of accessibility to wider market. Proper infrastructure, easy access to market and supply chain managementsare the more persistent concerns. The major opportunities for startups are in precision livestock farming, feed industry, smart AI and IoT based farming, e-marketing, farming processes and technology and so on. Sustainable livestock business is being practiced through various activities like diversification, value addition, precision farming, e-marketing, organic farming etc. Updating the technological interventions is the primary keyto success in market-oriented farming, as this will lead to meet the rising demand at all level.

Being the shock absorber of agriculture sector, livestock sector has diverse scope for startups and innovations. Based on a NASSCOM (2019), the country already has more than 450 agritechstartups, which is still increasing, @ 25% annually. At present there are 103 Livestock Tech startups in India (pashudhanpraharee.com). To support a more sustainable growth in livestock industry, Rashtriya Kamdhenu Aayog has beenformed, with an initial investment of Rs 500 crore. A funding of 60% from the Government will be given for Cow Dung and Urine Startups (https://krishijagran.com/). It is targeted to encourage more youth into initiating Cow based product Start Ups in India.

A few startups are doing great in livestock sector viz., moofarm, Stella APP, Licious, Fresh to home, Eggoz, Animall etc. They are using artificial intelligence and internet of things (IoT) based technologies as well as data science and analytics to foster an effective environment of farmers and entrepreneurs.

- **Licious:** Licious is an online platform to deliver of meat and seafood founded during 2015. Its app is available on Android and iOS platforms. It is India's first D2C Unicorn headquartered in Bengaluru, Karnataka. The company is operating on a farm-to-fork model and owns the whole back-end supply chain and cold chain. It is acting as One-Stop Platform for Fresh Meat & Sea Food. Abhay Hanjura and Vivek Gupta of IIT Delhi are the Cofounders.
- **Stellapps:** It provides end to end farm solutions for farmers, works on the concept of One-stop dairy supply chain digitization via IoT. IoT and data analytics technologies are used in order to improve Agri- supply chain parameters viz., milk production, procurement, cold chain, animal insurance, and payments.

Components of Stellapps

1. **mooON™:** To improve the productivity of cattle. It has 2 parts viz., moon device and moon app.

- **mooOn device:** It is a pedometer for cattle which helps to detect heat and various disorders. mooOn app is a herd management application to manage moon device.
- **SmartAMCU:** An automated Milk Colletion unit targeted to scale up business profits by effective milk procurement operation. It enables IoT-based, real-time acquisition and dissemination of milk procurement data.

2. **ConTrak:** Helps to regulate and control the cold chain to achieve better shelf life of milk during transport.

- **Animall:** It is an online trading platform for cattle & buffalo. Focused on building a peer-to-peer cattle trading platform. It has reached more than 8 million dairy farmers within 2 years duration. More than 850,000 cattle have been sold through this platform. It is a brainchild of Neetu Yadav and Kirti Jangra of IIT Delhi.
- **Eggoz:** It is a producer of eggs and supplies fresh eggs to households, local shops, and markets through online platform.
- **Activx Animal Health Technologies (Vetzz)**: It is a network of Veterinary Doctors engaged in providing immediate service to the animal owners. It works via real time tele consultation and provides doorstep visits (PIB, 2020).
- **Mooofarm**: It was created with an aim to bridge the gap between technology and dairy farmers. It has a built-in e-commerce platform for trading of animals.

Components

- E-Dairy Mitra: It helps to connect with an Expert Veterinarian, and to schedule a doorstep visit
- Digital Livestock Management: Gives platform to record cattle lifeline and get predictive analytics for farm
- E-commerce: Helps in purchasing high-quality dairy farming inputs
- MoooCoins: Encourages more use through exciting voucher and provides Moo coin on every data input. Targeted to attract more users.
- **Cowpathy**: Startup working with cow-based products which makes soap, hair oils, toothpaste, floor cleaners, incense sticks, shaving cream and face wash. It claims that the soap contains dried and pulverised cow dung, lavender powder,orange peel, and gooseberries and their toothpaste contain dung, ghee, and urine (https://krishijagran.com/).

Initiatives by Government of India to Promote Entrepreneurship Among Livestock Farmers

1. **Animal Husbandry Infrastructure Development Fund**: Department of Animal Husbandry and Dairying (DAHD), GoI has launched this Fund worth Rs. 15,000 crore as part of AtmaNirbhar Bharat Abhiyan Stimulus package for incentivizing investments by individual entrepreneurs, private companies, Farmers producers organizations (FPOs) etc. to establish infrastructure for dairy and meat processing and product diversification and animal feed plant.

2. **e-Gopala App** has been launched by Prime Minster of India to provide marketplace for farmers to trade quality germplasm, good quality cattle of high productivity. It is also providing AI services, vaccination and guiding farmers for animal nutrition and affordable quality treatment.
3. **Dairy Entrepreneurship Development Scheme:**DAHD implementing Dairy Entrepreneurship Development Scheme (DEDS) for creating self-employment opportunities in the dairy sector. It covers activities such as enhancement of milk production, procurement, and preservation, transportation, processing and marketing of milk. It is providing back ended capital subsidy for establishment of farms. The scheme is being implemented by NABARD.
4. **Animal Husbandry Startup Grand Challenge:** The DAHDandStart-up India, on partnership, launched this 'Animal Husbandry Start-up Grand Challenge'. It was planned to search for innovative and commercially reasonable solutions to address the six problem areas in prevailing in livestock sector. During 2019, it was launched by Hon'ble Prime Minister of India. A total of 12 startups were provided with cash grants worth Rs 1.02 Cr. The aim/challenge was to bring a unique solution for six problem statements. In 2021, the second phase of this challenge has been launched as **'Animal Husbandry Startup Grand Challenge 2.0:**The winners for each of the problem statements will be presented with Rs 10 Lakh cash prize whereas a runner-up will get Rs 7 Lakh cash (PIB, 2021).
5. **National Livestock Mission**: The Department of Animal Husbandry & Dairying, Government of India is implementing the scheme of National Livestock Mission with aims towards employment generation, entrepreneurship development, and increase in per animal productivity. The sub-mission on Breed Development of Livestock and Poultry has sharp focus on entrepreneurship development and breed improvement in poultry, sheep, goat and piggery by providing the incentives. NABARD is the funding agency.
6. **RKVY-RAFTAAR Scheme:** It is a unique scheme launched by Government of India, which is targeted to strengthen infrastructure in Agriculture and Allied sectors and to promote Agripreneurship and Agribusiness. It is facilitating financial assistance and fostering a system of business incubation. Till date, 24 RKVY-RAFTAAR Agribusiness Incubators (R-ABIs) throughout the country have been established. It has three components as given bellow:

- Agripreneurship Orientation: It is of 2 months duration. The individuals shortlisted will be provided with a monthly stipend of Rs. 10,000/-. Training is provided on various financial, technical, IP issues etc.

- Seed Stage Funding of R-ABI Incubatees: During this stage, funding upto Rs. 25 lakhs (85% grant from Government & 15% contribution from the incubatee) is provided.
- Idea stage/Pre-Seed Stage: Funding up to Rs. 5 lakhs (90% grant from Government and 10% contribution from the incubate/trainee) is provided during this stage (https://pib.gov.in/).

7. **KRI-TA-GYA 2.0: In the term KRI denotes Krishi (Agriculture), TA denotes Taknik (Technology) and GYAdenotes Gyan (Knowledge).** This unique event was organized with an aim to provide an opportunity to the students / entrepreneurs / innovators and others to display their innovative approaches &technology solutions in livestock farming. The theme was precision and economic animal farming. The former version of hackathon KRITAGYA was structured to promote potential technology solutions for improving farm mechanization. Special emphasis was given towards women friendly equipment.
8. **Innovation and Agri-entrepreneurship Development Programme**: It was launched under Rashtriya Krishi Vikas Yojana to promote innovation and agripreneurship. It is giving financial support for this. These startups are under various categories viz., agro-processing, artificial intelligence, waste to wealth, dairy, fisheries, farm mechanisation, etc.

Current Challenges in Startups in Agriculture and Allied Sectors

As per the reports by NABARD (2020), only 20% of agri-startups are profit-making. Around 44% of startups in agri-education are still looking for funds whereas majority of the startups possess a team size of less than 5 persons. Some other persistent hinderance are erratic Supply Chain, Involvement of Middlemen and improper access to credible information.

Proposed Solutions

- Focussing towards more commercially feasible solutions to the existing problem
- Keeping in view the current market status in livestock farming
- Streamlining the funding pattern of agencies
- Competency assessment of the aspiring entrepreneurs before approval of any start-and accordingly provision of refresher trainings
- Encouraging more ICT based solutions using IoT and Artificial intelligence to add transparency.

References

Agtech Funding Report: Year In Review 2016 | Agfunder.Com; Pp16

Background Paper, Promoting statups in Agriculture. Accessed on 23rd Mar. 2022 from http://nmoop.gov.in/conference/docs/Background_Paper_Agri_Startups.pdf

https://krishijagran.com/animal-husbandry/top-most-profitable-cow-startup-business-ideas/

https://thestartuplab.in/growth-story-of-indian-startup-ecosystem-and-top-5-most-promising-sectors-in-2021/

Krishnan, G. (2018). Indian Startups Are Coming of Age and This Is Just the Beginning. Retrieved from: https://inc42.com/resources/indian-startups-are-coming-of-age-and-this-is-just-the-beginning/

Mattoo, V. (2018). Meet the Top 5 Startups in India which aim at revamping the agriculture sector. Retrieved from: https://www.startuped.net/ebook/index.php/meet-the-top-5-agri-startups-helping-india-grow/

NASSCOM. Accessed on 23rd Mar. 2022 from https://nasscom.in/knowledge-center/publications/start-report-momentous-rise-indian-start-ecosystem

NASSCOM Report. Accessed on 23rd Mar. 2022 from https://nasscom.in/sites/default/files/media_pdf/NASSCOM_Press_Release_Agritech_Report_2019.pdf

Ohlan, M. and Saravanan, R, 2020. Discussion Paper 14. MANAGE- Centre for Agricultural Extension Innovations, Reforms, and Agripreneurship (CAEIRA)

PashudhanPraharee. Accessed on 23rd Mar. 2022 from https://www.pashudhanpraharee.com/top-livestock-agri-startups-innovation-for-boosting-the-future-of-livestock-agriculture-in-india/

PIB (2020) Accessed on 23rd Mar. 2022 from https://pib.gov.in/Press Release Page.aspx?PRID=1643788

PIB (2021) Accessed on 23rd Mar. 2022 from https://pib.gov.in/PressReleaseIframePage.aspx?PRID=1781031

Shagun, 2021, Agri share in GDP hit 20% after 17 years: Economic Survey. Down to earth. Accessed fromhttps://www.downtoearth.org.in/news/agriculture/agri-share-in-gdp-hit-20-after-17-years-economic-survey-

StartUp India Website. Accessed on 23rd Mar, 2022 from https://www.startupindia.gov.in/content/sih/en/international/go-to-market-guide/indian-startup-ecosystem.html

5

Startup Opportunities in Veterinary Sector

Sanku Borkataki[1] and Pankaj Goswami[2]

[1]Associate Professor, Division of Veterinary Parasitology, Faculty of Veterinary Sciences and Animal Husbandry, R.S. Pura, Sher-e-Kashmir University of Agricultural Sciences and Technology, Jammu

[2]Professor, Division of Veterinary Pathology, Faculty of Veterinary Sciences and Animal Husbandry , R.S. Pura, Sher-e-Kashmir University of Agricultural Sciences and Technology, Jammu

Unemployment is a big challenge in every sector. In Jammu and Kashmir unemployment rate is 22.2 percent as per the Centre for Monitoring Indian Economy (CMIE) figures, and the rate is even higher than the National level unemployment rate of 7.1 percent recorded across India. Livestock sector plays an important role in rural livelihood, employment and income generation. There is a huge opportunity in this sector for developing start-up at their own and generation of employment. This is all possible if the skilled graduate of animal husbandry can be motivated for fostering new idea of startup which is market driven. Major livestock products like milk and milk products, meat and eggs contribute around one-sixth of the calories and one-third of the proteins in the per capita food supplies of the world. Animal husbandry is the second largest economic activity of rural India. Animal husbandry is a growth engine and annual growth rate in dairying is 5 % and in poultry it is 10% and will alleviate rural poverty and uplift the rural farmer. The availability of per capita animal protein is 10.8g whereas the requirement as per world average is 25g. National sample survey reports that 70-75% of their food budget is for milk and milk products. The need of the hour is increase in production, productivity and improvement through value addition in the marketing channel. Commercialization of cow dung and urine will also encourage people to not discard cows which have stopped milk production. By-products like urine and dung can be used for medicinal and agriculture purposes. It has been reported that startups would receive upto 60 per cent initial investment from Government for by-products like cow dung and urine. Eco friendly products can be prepared from dung and urine like Bio CNG, vermicompost, incense sticks, flowerpots, dung log etc. Cow dung soaps, shampoos, toothpaste, shaving creams, sunscreens, face wash, incense sticks and biofertilizers are the major dung and urine products which are present in the market and now even marketed by many online websites like Amazone, Flipkart etc, since livestock itself is a sustainable business, whosoever will get engaged in this form of startup business, that will definitely have edge to prosper, as such no activities is going on in UT J&K at a large scale.

India is emerging as a global destination for Innovation, Incubation and Start-up. Government of India has come up with exciting programme to support startup ecosystem and to nurture innovations and has launched Startup India action Plan by bringing in policies and frameworks that can support and nurture the start up eco-system. NITI Aayog has developed a collection of tips and schemes which will support India's ability within the international stage, by supporting the Nation's innovators. India ranks third in Start up ecosystem after United States and UK and is emerging as a hub for innovation and has a large population of youth with untapped ideas. Innovations can trigger social change and solve problems faced by contemporary India. Recent findings reveal that success rate in startups are only 30 percent. International trading system, economic impact, policies, technology solutions and digital innovations are the major areas which facilitates Startups in the country. Today youth is emerging as job creators rather than job seekers.

Some of the Important Thematic Areas are Given Below Where Specific Startups are Coming with Multiple Innovative Ideas with Technological Solution

i. Traditional Knowledge practices in veterinary care

ii. Wealth from livestock waste

iii. Sensors for management and health of livestock

iv. Artificial intelligent system/tool for optimum productivity, better health and welfare of animals

v. Value addition of milk and milk products

vi. Climate smart dairy farming

vii. Farm data management and sharing in livestock farming

viii. Dairy farm mechanization automation and robotization

ix. Nutrition, breeding, management and diagnostic interventions for livestock

x. Technology options for relief and rescue of animals in Disaster

xi. Innovative livestock business models

1. Traditional Knowledge Practices in Veterinary Care

Ethno veterinary practices include the use of local medicinal plants to prevent, cure or treat various ailments in animals. Ethno veterinary practices are known in India since the ancient Vedic period. Ethno veterinary practices concern to animal healthcare is a century old practice stated back to domestication of various livestock species. It involves use of domestic knowledge, belief, practices and skills pertaining to healthcare and management of human, animals and birds. This branch of science, known as ethno veterinary practices or EVPs is becoming a growing concern in the field of ethno botany. India being a developing country with rich ethno veterinary knowledge, is sought to be a 'EVP hub' owing to practice with decades of experiences. Since time immemorial, the practice is mainly based on the use of plant formulations and other locally available cheap ingredients. Livestock raisers and local people with a strong knowledge of veterinary medicine usually follow traditional ways of classifying, diagnosing, preventing and treating common animal ailments. The traditional medicines that are commonly used for animal healthcare can cut down costs considerably. Moreover, they are readily available to the ordinary farmer. It can be a major alternative to antibiotics use for the purpose of promoting growth and production performance or for prophylaxis and treatment of various pathogenic organisms.

Jammu and Kashmir is blessed with diversified plants with medicinal properties. Exploration of these medicinal value in the plant and it identification bio molecule can be used for various therapeutic preparations. Thus start-up in this sector is very much promising to fill the needs of traditional value in therapeutics as well as conservation of germplasm of medicinal plants.

2. Opportunities on Waste Management in Animal Husbandry Practices

Cow dung processing

Cow dung is a unique fertilizer having rich source of nutrient for growth of plant. The cow dung can be processed for making various useful purposes. Many major businesses have started to deal with cow dung and cow urine based products. Some of them have even made a million dollar empire. All these products have been completely based on pure extracts of cow urine and cow dung. Cow dung soaps, shampoos, toothpaste, shaving creams, sunscreens, face washes, teas, incense sticks and bio fertilisers are the major cow dung and cow urine products which are present in the market and are now even marketed by many online websites like Amazon, flip kart etc. As long as the smell and the quality are concerned, the reviews of these products online have been quite good. People claim of these products smelling good and also providing extra care to their skin. One more factor that plays an important role in the quality of these products are the cow varieties that are raised with immense care and are provided with proper nutrition. Like all of the companies stick to a particular variety of cow breed specially the Indian species claiming that the cow dung and cow urine have special properties when fed on special grass or food. Now at present there are many more companies which deal with such products even if not pure cow urine and cow dung but a mixture of it too. Panchagavya products like the products made up of the five products from cow like milk, ghee, cow dung, cow urine and curd, have been produced and marketed online and offline both from the past years too. In Madhypradesh, cow dung products such as manure, cow dung cakes (used as fuel), diyas (lamps) and flower pots made by more than 4,000 women of 354 self-help groups (SHGs) and are now available on e-commerce platforms which is very much encouraging. Earlier, manure was being sold only at the cowsheds, but now it is selling online on platforms such as Amazon due to the increasing demand for manure from public. Thousands of rural and poor women engaged in making products from cow dung in Chhattisgarh's Rajnandgaon district have received a fillip with the administration now ensuring that their products reach a wider market through e-commerce platforms, following a rising demand including from other states.

Panchagavya Therapy

In Indian culture, Cow is the backbone of Indian culture and rural economy, and sustains our life, represent cattle wealth and bio-diversity. Cow is known as "Kamdhenu" and "Gaumata" because of its nourishing nature like mother. Panchgavya is a term used to describe five major substances, obtained from cow, which include cow's urine, milk, ghee, curd and dung. All the five products possess medicinal properties against many disorders and are used for the medicinal purpose singly or in combination withsome other herbs. This kind of treatment is called Panchgavya therapy or cowpathy. all five products can be used alone or combined or any other synthetic, herbal, or mineral origin. Panchgavya therapy is recommended for a variety of diseases viz., asthma, flu, allergies, cardiovascular diseases, renal disorders, rheumatoid arthritis, leucoderma, wound healing, leucorrhoea, hepatitis, dietary and gastrointestinal tract disorders, obesity, tuberculosis, ulcer, chemical intoxication, other bacterial, fungal and viral infections. The therapy has also demonstrated its therapeutic potential against severe pathological conditions like cancer, AIDS, and diabetes

3. Startup on Livestock Product Technology Innovation

Value addition of milk and milk product

Value addition can be defined as a process of increasing economic value and economic appeal of a commodity. Value is added by changing a commodity's form, colour, taste and other such methods to increase the shelf life of perishables. Value of a product can also be added by capturing the market at the right time. This may include transporting the product to places where it can earn more income, or storing it and selling when there is high demand. A change in the physical state or form of the products and production of a product in a manner that enhances its value as demonstrated through business plan. Value addition minimizes wastage and improves quality of a commodity which realizes better prices.

Jammu and Kashmir is ideally suited for dairy development. The state has registered a steady growth in milk production. The state's milk production in 1995-96 was 3.69 lakh metric tonne which stood at 25.1lakh metric tonne in 2019-20. In 2000-2001, Jammu and Kashmir became a milk surplus state. Dairy farming is a household activity largely done by domestic labour in Jammu and Kashmir. A large number of women are mainly engaged constituting a major percent of the total agricultural labour force and the majority of families have two to three domestic labourers. This shows that in Jammu and Kashmir, dairy development activity is still a subsistence activity but income and employment opportunities can be increased through the modernization of this sector. The Jammu district, which is contiguous to Poonch and Rajouri districts where Gujjar population predominates, is ideally suited for dairy development.

The dairy sector in Jammu provides many opportunities for youth to operate micro and small-scale enterprises. This is because milk is part of farm produce that generates cash on a regular basis and it is one of the foods consumed by almost people of all ages in good quantities. Micro- and small-scale enterprises are affordable and manageable by rural people. They create a large number of non-farm employments and income opportunities in relatively poorly developed areas and require small capital and little sophisticated managerial and technical skills. However dairy farmers in Jammu mostly sell raw milk, which fetch low prices for their produce. Most of the families in tehsil Ranbir singh Pura of the Jammu division are engaged with dairy farming and a large number of women are mainly engaged in various dairy activities. They transport their produce more than 20 Km away from the site of production to the Jammu city as there are very few processing units in the area and fetch meager amount for their produce. One-way dairy farmers can improve this situation involves value addition on raw milk. Value addition minimizes wastage and improves quality of a commodity which realizes better prices.

Motivation of young youth and capacity building in the area of milk processing may contribute poverty reduction through entrepreneurship development with a particular focus on rural people/unemployed youth. The essential elements are to incorporate the technical skills that encourage the initiatives of entrepreneurs and to enhance the human capacities required to foster entrepreneurial dynamism. To respond to the needs of unemployed youth and to materialize their economic potential and thereby to improve their standard of living, it is necessary to motivate the budding entrepreneurs.

4. Artificial Intelligence in Animal Health

Artificial intelligence is an recent emerging field for technocrats to develop startup with sensor mechanics, ray based technique ets. Biosensors, as an application for animal health management, are an emerging market that is quickly gaining recognition in the global market. Globally, a number of sensors being produced for animal health management are at various stages of commercialization. Some technologies for producing an accurate health status and disease diagnosis are applicable only for humans, with few modifications or testing in animal models. Now, these innovative technologies are being considered for their future use in livestock development and welfare. Precision livestock farming techniques, which include a wide span of technologies, are being applied, along with advanced technologies like microfluidics, sound analyzers, image-detection techniques, sweat and salivary sensing, serodiagnosis, and others.

6

Prospects and Potential of Agri-Business Entrepreneurship for Enhancing Income and Employment: Success Stories

J.P. Sharma and Pawan Kumar Sharma[1]*

**Former Vice Chancellor, Sher-e-Kashmir University of Agricultural Sciences and Technology, Jammu*

[1]Assistant Professor, Directorate of Extension Education, Sher-e-Kashmir University of Agricultural Sciences and Technology, Jammu

India is an agrarian country and the agriculture and allied sector is the largest employer of workforce as it accounted for a sizeable 18.8 per cent in 2021-22 in Gross Value Added (GVA) of the country (Economic Survey 2021-22).Agriculture is the primary source of livelihood for about 58% of India's population and only the food processing industry employs about 1.77 million people (Indian Agriculture and Allied Industries Industry Report, November 2021). In the year 2019-20, gross value added by agriculture, forestry, and fishing was estimated at Rs. 19.48 lakh crore (US$ 276.37 billion). But still farming community has to face lot of hardships. The farmers are still applying chemical fertilizers on the basis of their own experiences as they don't possess the soil heath cards. Although central government has initiated Soil Health Management (SHM) scheme to assist State Governments to set up new static Soil Testing Laboratories (STLs) and Mini Soil testing Labs (MSTLs). But it is time consuming and cumbersome process. There is great opportunity for Agri-business entrepreneurs with a commercially viable solution to farmers by development of simplified, sensor based and quick soil testing methods to test the nutrients and micronutrients.

Although India has achieved self-sufficiency in food grain production, but still one of the biggest issues facing the agricultural sector in India is low yield. The average productivity of different food grain crops in India is 30-50% lower than that of developed nations. The major factors contributed in low productivity includes: average farm size, poor infrastructure, low use of farm technologies and best farming techniques, decrease of soil fertility due to imbalance application of fertilizers and pesticides, limited access to credit, support expertise, educational services or irrigation solutions, Poor storage facilities. The challenge of seeking affordable, accessible, easy-to-use technologies, products or services to enhance agricultural productivity in India creates opportunities for young entrepreneurs to establish their startups to provide timely solutions. More than 40% of food production is destroyed by insect-pests every year. This loss occurs despite of the application of approximately 3 million tons of pesticide per year plus the use of a wide array of non-chemical controls, like crop rotations and biological control. Due to lack of effective, affordable and eco-friendly technologies to control pests, farmers are left with no choice but to continue spraying harmful and toxic pesticides on crops. The challenge creates new opportunities for young minds to come up with technology solutions to substitute the use of pesticides & insecticides to prevent pre-harvest losses. Burning of crop stubbles, stray and other left over contributing between 12 and 60 Particulate Matter concentration and has been identified as a major health hazard. In addition, it causes loss of vital components such as nitrogen, phosphorus, sulphur and potassium from the top most layer of the soil making the land, less fertile and unviable for agriculture in the long run. These challenges are creating new opportunities for young agripreneurs to come out with technological solutions to discourage crop residue burning. To do the agriculture on scientific lines requires availability of small agricultural implements/ micronutrients/certified quality seeds. This creates scope for youths to come out with agri-business entrepreneurs that can provide small agricultural machinery to farmers through Custom Hiring Centres and other necessary inputs through online/call center.

The concept of Agri-Business is now changing minds and farmers' also are now becoming an entrepreneur. Small-scale farmers all over the world have shown a remarkable ability to adapt. They look for better ways to organize their farms and enhance their income. They try new crops and cultivars, better animals, and alternative technologies to increase productivity, diversify production, and reduce risk – and to increase profits. They have become more market oriented and have learned to take calculated risks to open or create new markets for their products. They need to be innovative, forward-looking, capability of managing their businesses as long-term ventures and ability to identify opportunities and seize them. The farmer having an entrepreneurial spirit try to act differently and make decisions about the value chain to be followed in his farm to increase the return and profits of the farm business.

Agribusiness is a term used to describe the sector that encompasses all economic activities that are related to farming, i.e., breeding, crop production/farming, farm machinery, distribution, marketing etc. It relates to industries that are engaged in farming or that produce farm inputs. Examples of agribusiness include farm machinery manufacturing, seed supply, and agrichemicals. It also is used to describe businesses that are involved in the marketing and distribution of farm products. These businesses include warehouses, wholesalers, processors, retailers, and more. Any company that participates in the production, marketing, safety, and distribution of food is involved in agribusiness. More precisely it is defined as a sector that supports the growth of the agricultural industry, which is pivotal to economic growth. It also continues to play a crucial role in the growth of developing countries. Agribusinesses can potentially improve agricultural productivity, which is why governments often offer subsidies to agricultural businesses. Agricultural activities also contribute to an improved system of food security and sustainable food production, as well as income for a majority of the poor in developing countries. However, the activities increase the emission of greenhouse gases and contribute to global warming – which is why innovation is important in the sector to address such problems.

Entrepreneur refers to an individual that has an idea and intends to execute on that idea, usually to disrupt the current market with a new product or service. Entrepreneurship in agriculture can also be defined as the formation of novel economic organization for the intention of growth under risk and uncertainty in agriculture (Dollinger, 2003). Contrary, Gray (2002) defines an entrepreneur as an individual who controls a business with the purpose of growing the business along with leadership and managerial skills necessary for achieving those goals. The most prominent example of entrepreneurship is the starting of new businesses. In economics, entrepreneurship connected with land, labour, natural resources and capital can generate a profit. The entrepreneurial vision is defined by discovery and risk-taking and is an indispensable part of a nation's capacity to succeed in an ever-changing and more competitive global marketplace. Many small-scale farmers and extension organizations understand that there is little future for farmers unless they become more entrepreneurial in the way they run their farms. They must increasingly produce for markets and for profits. Becoming more entrepreneurial can be a challenge for small-scale farmers. They will need help from extension workers and other institutions.

Importance of Agri-Business Entrepreneurship

Agri-Business entrepreneurship is important for national economy in following ways (Sah, 2009).

i. It helps in achieving productivity profit by small farmers and amalgamating them into local, national and international markets.

ii. It helps in decreasing food costs and provides high quality diets to the rural and urban poor in the country.

iii. It accelerates growth, diversifies income and develops entrepreneurial opportunities in both rural and urban areas.

Prospects and Potential

In India, the losses in postharvest sector are estimated to be from 10 to 25 per cent in durable, semi-perishable products like milk, meat, fish and eggs. The estimated losses in fruits and vegetables are higher and reached from 30 to 40 per cent. These percentages are not acceptable and adversely affect the Indian economy. Only 2.2 % of commodities are processed. In contrast, countries like USA (65%) and China (23%) are far ahead of India in reducing the post-harvest losses and enhancing the value addition and shelf life of the farm products. This has ample scope to provide technological solutions to increase the efficiency of the agricultural chain and ultimately reduce waste while increase farmers' earnings. The potential and scope for agri-business entrepreneurship in India can be summarized as under:

1. India has diverse climatic conditions for production of various types and varieties of fruits and vegetables such as tropical, subtropical.
2. There is growing demand for agricultural inputs like hybrid seeds, fodder, organic fertilizers etc.
3. Export can be harnessed as a source of economic growth. As a signatory of World Trade Organization, India has vast potential to improve it present position in the World trade of agricultural commodities both raw and processed form. The products line include cereals, pulses, oilseeds and oils, oil meal, spices and condiments, fruits and vegetables, flowers, medicinal plants and essential oils, agricultural advisory services, agricultural tools and implements, meat, milk and milk products, fish and fish products, ornamental fish, forest by products etc (Sundar, 2016)
4. Production of vegetables and flowers under green house conditions can be taken up to harness the export market.
5. Trained human resources in agriculture and allied sciences will take on agricultural extension system due to dwindling resources of state finance and downsizing the present government agricultural extension staff as consulting services.
6. Development in agriculture increases the employment opportunities in the country
7. Livestock in the country can provide one of the successful enterprises

8. Bee keeping and apiary is one of the profitable enterprise.
9. As India is rich in forest resources, it provides opportunities in this sector also.

Agribusiness Entrepreneurship Potential in India

India has various areas where agribusiness is being done successfully. Organic farming, fruit juices, fruit pulps, flavours, candies, pickled products, frozen food items, processed milk products, value added products, organised food retailing, mushrooms, medicinal plants are few of them. They are discussed below in detail:

1. **Organic farming:** Organic agriculture is expanding fast. The share of agricultural land and farms continues to grow in many countries. 2.4 million producers were reported, more than 3 quarters are in developing countries. Asia and Africa constitute for more than half of world organic producers with 35% and 30% producers respectively followed by Latin America 19%, Europe 14%, North America 1% and Oceania/ Australia 1%. Oceania/ Australia constitutes for only 1% of producers despite having the maximum share of land which shows that Australia has highest per capita land holding. Coming down to countries, India has highest number of organic producers followed by Ethiopia, Mexico and Uganda (FiBL, 2017). Organic agriculture market contributes to US $ 81.6 billion. The global organic food market which is estimated at USD 90 billion in 2015 has been growing at a CAGR of around 12% of last 14 years. In 2016, the global organic food market has a value of USD 100 billion, an increase of 60% since 2010.Europe and North America together generate about 90% of the global organic food sales. USA, Germany, France and Italy are the leaders in consumption of packaged organic products. United States leads the market with a market size of over USD 39 billion (47%) followed by Germany (11%) and France (7%). Japan leads the Asian countries in terms of organic food consumption with nearly 54% of the share in 2010 (IFOAM, 2017). In terms of highest per capita consumption, Switzerland tops the chart with 262 Euros followed by Denmark 191 Euros and Sweden 177 Euros in 2015 (FiBL, 2017).

   India has a total of 12.07 lakh ha area under organic management out of which 6.40 lakh ha area is organically certified and 5.67 lakh ha is under conversion in 2012-13 (NCOF Ghaziabad, 2014). It has been increased from 42 thousand ha in 2003-04 to 12.07 lakh ha with a CAGR of 45.22 percent (APEDA, 2015). In India, Orissa has maximum OA land of 95.7k ha followed by Gujarat (70.5k ha), Uttar Pradesh (53.5k ha) and Karnataka (51.5k ha). In terms of producers, Madhya Pradesh has maximum farmers (i.e. 1.77 lakh) followed by Maharashtra (65.6k), Orissa (62.1k) and Uttrakhand (47.1k). Sikkim is the fully organic certified state of India. The maximum OA land was under cotton cultivation (3.53 lakh ha) followed by oilseeds excluding Soybean (1.79 lakh ha) and fruits and vegetables (1.70 lakh ha). India has a total of 5.98 lakh farmers under organic certification process out of which 3.51 lakh are certified organic farmers while 2.47 lakh are in conversion process. 2099 operators, 427 processors, 919 grower groups, 253 exporters have been reported in 2009. Total quantity of organic produce was 1811111 MT out of which 53918 MT was exported. The total value of exports were 116.09 millions $ or `591 Cr (NCOF Ghaziabad, 2014). Organic agriculture market of India contributed to `2500 Cr in 2014-15 which is lesser than even one percent of global organic food market out of which largest share was of medicinal and herbal plants (45%) followed by cotton (15%), oilseeds (9.30%) and vegetables and fruits (9%). Rice alone constitutes for 4.50 percent share of OA market. Organic food market in India can be segmented into domestic market and export market which made a share of 31 percent and 69 percent in 2013-14. The OA market of India has been continuously growing from Rs. 675 Cr in 2009-10 to Rs. 1928 Cr in 2013-14 at a CAGR of 30 percent (Anoymous, 2015). According to Mukherjee et al. (2014), a large number of companies are investing in organic business activities due to increasing demand for organic food. Maximum companies are engaged in farming and manufacturing activities and minimum companies are engaged in import activities.

2. **Indian Organised Retail Market:** Organised retailing refers to trading activities undertaken by licensed retailers, that is, those who are registered for sales tax, income tax, etc. These include the corporate-backed hypermarkets and retail chains, and also the privately owned large retail businesses. In other words, it is a network of similarly branded stores with an element of self-service. Organised retail in India today holds only a fraction of the market share potential in India. In 2001-02, organized retail trade in India was worth Rs 11,228.7 billion. It has risen from 0 to 6 per cent in a very short period mainly on volumes and not a value-driven growth (Gupta, 2012). (According to McKinsey & Company report titled 'The Great Indian Bazaar: Organised Retail Comes of Age in India', organised retail in India is expected to increase from 5 per cent of the total market in 2008-09 to 14-18 per cent of the total retail market and reach US$ 450 billion by 2015-16. The food and grocery constitutes the highest retail volume and this share has shown a tremendous growth over the years. According to NSSO 60th round, 54 per cent of the rural and 42

per cent of urban expenditure was on food. The second largest share is commanded by the apparels. Clothing and textile is the largest organised market and is dominated by retailers like Pantaloon, Westside, Globus, Koutons. This owes to the increasing disposable incomes and changing lifestyles. Industry trends for retail sector indicate that organized retailing has major impact in controlling inflation because large organized retailers are able to buy directly from producers at most competitive prices.

3. **Hydroponics:** Hydroponics is the cultivation of plants by placing the roots in liquid nutrient solutions rather than in soil (soilless growth of plants and without conventional soil). It uses artificial means to support and nutrient solutions to provide the required nutrition to plants. A person interested in such type of technology can start his business with hydroponic retail store. Moreover, this area is gaining momentum from the last few years.
4. **Vermicompost Production:** Vermicomposting is a simple biotechnological process of composting, in which certain species of earthworms are used to enhance the process of waste conversion and produce a better product. It has become major component of agri-business models with low investment cost.
5. **Hardening Nurseries:** Hardening of the plants in the nursery includes "Any treatment that makes the tissues firm to withstand unfavourable environment like low temperature, high temperature and hot dry wind". Such type of business is also gaining importance as far as agribusiness is concerned.
6. **Processing Industry:** Fruit and vegetable processing, including dehydration, canning, aseptic packaging, packing, processing of underutilized fruits, and processing for other products walnut, grape raisin, air-dried fruits, fruit candies and spices' powders etc. Processing of maize for starch and feed through improved mini/ small mills and dry milling plants, processing of sugarcane for various jaggery products like spiced jaggery, powdered jaggery, and jaggery cubes, processing of dairy products and many more are part of the processing industry. As demand for processed foods is increasing day by day so the opportunity for agribusiness is also increasing because all the processing industries falls under agribusiness ventures.
7. **Milk Cooperatives:** Now a days selling of milk on cooperative basis also forms the part of agribusiness model. Dairy farmers constitute a group for collection of milk at their own. The milk is than passed through chilling process at collection centre and from the centre it is transported to cooperatives for further processing of the milk.

8. **Packing and Packaging:** Package attracts consumer's attention to particular brand, enhances its image, and influences consumer's perceptions about product (Rundh, 2005). Also package imparts unique value to products (Underwood et al., 2001; Silayoi and Speece, 2004), works as a tool for differentiation, i.e. helps consumers to choose the product from wide range of similar products, stimulates customers buying behavior (Wells et al., 2007). Thus package performs an important role in agribusiness and could be treated as one of the most important factors influencing consumer's purchase decision. In this context, seeking to maximize the effectiveness of package in a buying place, the researches of package, its elements and their impact on consumer's buying behavior as far as agribusiness is concerned became a relevant issue. A packaging that is well designed for its marketing function helps to sell the product by attracting the attention and expressing the product positively to the ultimate potential consumers. To achieve the communication goals effectively and to optimize the potential of the packaging in a buying place, marketers must recognize the attributes of a packaging and their impact on consumer's purchase decision and integrate these perceptual attributes into their packaging design.

9. **Fast Moving Consumer Goods:** Fast Moving consumer goods are those goods that are consumed every day by the average consumer and are replaced or fully used up over a short period of days, weeks, or months, and within one year. As per the report of Indian brand equity foundation (2015), the Indian FMCG sector is the fourth largest in the economy and has a market size of US$13.1 billion. The overall FMCG market is expected to increase at a compound annual growth rate (CAGR) of 14.7 per cent to touch US$ 110.4 billion during 2015-2020 with the rural FMCG market anticipated to increase at a CAGR of 17.7 per cent to reach US$ 100 billion during 2015-2025. The foods category in FMCG is gaining popularity with a swing of launches by HUL, ITC, Godrej, and others. As per the report of Indian Brand Equity Foundation (2015), this category has 18 major brands, aggregating Rs.4,637crore. Amul India's largest food company has a good presence in the food category with its ice-creams, curd, milk, butter, cheese, and so on. Britannia also ranks in the top 100 FMCG brands, dominates the biscuits category and has launched a series of products at various prices.

10. **Floriculture:** Floriculture is now profitable venture worldwide as it is one of the fastest growing crop trends based on its demand. People are using flowers in their day to day activities also; moreover, it is mostly used in

marriages and functions as loose flowers. Demand for cut dry flowers has also increased tremendously for so many years.

11. **Export of Fruits and Vegetables:** Now-a-days government is providing support to the farmers for exporting their product to get better price for their produce. Farmers' can start their own enterprise/ business for export of good quality fruits and vegetables where demand for the same exists.
12. **Frozen Foods:** Due to the paucity of time, demand for frozen food items is increasing continuously. An entrepreneur interesting in such type of business will definitely earn a handsome income.
13. **Mushroom Production:** Mushroom production business can be a mean of big profit in just a few weeks with considerable low investment to start a business. Button mushroom is demandable now a days. Oyster mushroom is also important as far as the environment and temperature of Jammu is concerned. Post harvesting of the same is very easy. Oyster mushroom is also used in the pharmaceutical industry for boosting of immune system and control of diabetes. Soap and detergent companies are also using it in their products. We can form powder of oyster mushroom after drying it and same can be consumed with milk.
14. **Juice-Jam-Jelly Processing Industry:** Processing of fruits into juice, jam and jelly has the huge business opportunities as these products fall under ready to drink or ready to eat products category. The demand for these products is at its peak in the current time and so the business opportunity is also high.

Besides above agribusiness opportunities discussed in detail, some of the agro based industries includes Flour milling, groundnut processing, cashew-nut processing, snail farming, sunflower farming, bee-keeping, retailing, organised retailing, fish farming etc.

Enhancing Income and Employment

Increase in Farmer income is a big task and Government of India is very particular to enhance it through different means. Few procedures are much needed to adapt in the form of models to enhance the income of farmers and these are as:

Promoting Innovations: Innovations are need of the hour to grow as entrepreneur or start-up depending upon the solution to the problems. We need to promote innovations through farmers and young entrepreneurs as these needs to participate actively in designing, testing, adapting and introducing innovations to the agriculture.

Increase in Value Addition: Farmer-entrepreneurs need to realize that capturing value requires producing for buyers and final consumers. But just producing and selling is insufficient. This requires greater understanding and knowledge of value chains and their different elements. It requires a plan for participating further down the value chain. Farmers capture more valueby selling their regular fresh produce directly to consumers. Another way is to sell high-value products such as organic products. Instead of producing commodities that are undifferentiated, farmers need to produce commodities that are differentiated such as organic fruit, a unique type of commodity or other specialized products for a limited "niche" market. Another way of adding value is to enter into production and marketing contracts. Contracts are increasingly being used between farmers and input suppliers, farmers and processors and between farmers and other buyers for the production and supply of fresh or processed agricultural produce.

Development and Adaptation of New Technologies: Farmers need to be encouraged to develop and adapt new technologies and share them with other farmers. Extension workers can facilitate partnerships among farmers and between farmers and researchers to work together to identify, develop and test new technologies and practices to improve productivity and profitability.

Use of Information Technology: Now-a-days Information Technology is available and accessible everywhere. Therefore, it is a very powerful way for extension workers to educate and inform farmers about new ideas, technologies and other information. Physical distances and the lack of transportation facilities often limit the ability of extension workers to share information with farmers. Mobile phones, tablets, and computer-based systems can be used to overcome these physical barriers. The challenge is how information technology can be harnessed for the benefit of both extension agents and farmers without compromising the importance of human and unique local factors.

Role of Farmer Producer Organizations

An Farmer Producer Organization (FPO), formed by a group of farm producers, is a registered body with producers as shareholders in the organisation. It deals with business activities related to the farm produce and it works for the benefit of the member producers. It provides end-to-end support and services to the small farmers, and cover technical services, marketing, processing, and others aspects of cultivation inputs.

The year 2014 was observed as the "Year of Farmer Producer Organisations **(FPO)**" by the Government of **India**. As per the concept, farmers, who are the producers of agricultural products, can form groups and register themselves under the **Indian** Companies Act. Presently, around 6000 FPOs (including

FPCs) are in existence in the country, which were formed under various initiatives of the Govt. of India (including SFAC), State Governments, NABARD and other organizations over the last 8-10 years. Govt of India has already taken an initiative to form 10000 FPOs by 2024.

Benefits of Farmers' Collectives

- Some of the important benefits of organizing farmers into Collectives, as demonstrated through various pilots, are as under: Cost of production can be reduced by procuring all necessary inputs in bulk at wholesale rates
- Aggregation of produce and bulk transport reduces marketing cost, thus, enhancing net income of the producer
- Building the scale through produce aggregation enables to take advantage of economies of scale and attracts traders to collect produce at farm gate
- Access to modern technologies, facilitation of capacity building, extension and training on production technologies and ensuring traceability of agriculture produce
- Post-harvest losses can be minimized through value addition and efficient management of value chain
- Regular supply of produce and quality control is possible through proper planning and management
- Price fluctuation can be managed; if there are practices like contract farming, agreements, etc
- Easy in communication for dissemination of information about price, volume and other farming related advisories
- Access to financial resources against the stock, without collaterals
- Easy access of funds and other support services by the government / donors / service providers, and
- Improved bargaining power and social capital building.

FPO Service Model

The FPO offers a variety of services to its members and helps to increase income and employment directly and indirectly. It can be noted that it is providing almost end-to-end services to its members, covering almost all aspects of cultivation (from inputs, technical services to processing and marketing). The FPO facilitates linkages between farmers, processors, traders, and retailers to coordinate supply and demand and to access key business development services such as market information, input supplies, and transport services. Based on the emerging needs, the FPO will keep on adding new services from time to time. The set of services include Financial, Business and Welfare services.

Entrepreneurship Development Opportunities in Agriculture

Agribusiness increases the scenario and opportunities for value addition, packaging, branding, retailing, and exports of agricultural commodities through improved technology and agribusiness management. Indian population directly or indirectly depends on agriculture and now agribusiness is taking the lead and raw material for industries comes from agricultural area only. Now, the time is approaching where agribusiness is most likely to control the Indian economic growth. Presently, there are about 25 per cent to 30 per cent of post harvest losses in fruits andvegetables before reaching the final consumer i.e., due to lackof proper post harvest management practices. However, introduction of agribusiness in agriculture has created lot of changes in business professionalism. The scope and opportunity in the agribusinesshas also been increased. The possible opportunities of entrepreneurship in agriculture are:

i. Agro produce processing units – New product is not manufactured but the agricultural produce is processed to form end product. rice mills and dal mills are few examples.

ii. Agro Produce manufacturing units – In this category, new productsare produced depending upon the agricultural produce as the main raw material.sugar factories and bakery are few examples. .

iii. Agro-inputs manufacturing units – These units are engaged in the production of goods either for mechanization of agriculture or for increasingagricultural productivity.Fertilizer and pesticide production units, foodprocessing units and agricultural implements are few examples.

iv. Agro service centres – Agricultural engineering workshops and servicecentre for repairing, maintenance and serving the agricultural implement used inagriculture are the part of agro-service centres.

Bansal (2011) has Proposed Some Entrepreneurial Areas in Agriculture

1. Farming- Crop, dairy/poultry/goat, fish, rabbit, vegetables, flowers, ornamental plants, palmrosa, fodder, sericulture, agro-forestry, beekeeping, mushroom.
2. Product Marketing- Wholesale, retail, commission agent, transport,export, finance, storage, consultancy.

3. Inputs Marketing- Fertilizer, agricultural chemicals, seeds,machineries, animal feed, poultry hatchery, vet-medicines,landscaping, agriculture credit, custom service, bio-control units,bio-tech units.
4. Processing- Milk, fruits, vegetables, paddy, sugarcane, cashew, coir, poultry, cattle, tannery, brewery.
5. Facilitative- Research and development, marketing information,quality control, insurance, energy.

Opportunities for Agri-Business Entrepreneurship in India

Agri-business entrepreneurship is the future of India and Government of India is already focusing on it. Farmers and agri start-ups willing to grow are having number of opportunities for agri-business entrepreneurship and some of them are listed below:

i. Input industry is one of the important industry where demand for agricultural inputs have increased.
ii. Agricultural Biotechnology is having scope in the form of seed production, bio-control agents, industrial harnessing of microbes fordifferent products.
iii. Processing is another area where agri-preneurs are having opportunities and can excel.
iv. Animal husbandry as an industry can become a huge source of generating employment and livelihood. Introduction of new product (Designer Egg, Crossbred cows, Hybrid fowls, etc.) are some of the areas where one can excel as agripreneur.
v. Value addition to the livestock products such as milk, egg, meat, and fish have huge profit potential. Value of the products get increased many folds during processing, and thereby provide excellent returns.
vi. Non-Timber Forest Products are unexploited and therefore is having scope for developing an enterprise.
vii. Beekeeping, apiary and mushroom production can be taken up on great extent in India.
viii. Organic farming has huge potential in India as the pesticideand inorganic fertilizer application are less in India compared toindustrial nations of the world.
ix. Huge opportunities for production and promotion of bio-pesticidesand bio-control agents for protection of crops.
x. Micro-irrigation systems and labour saving farm equipments havepotential in future due to declining groundwater level and labour scarcity for various agricultural operations.

xi. Green House production of vegetables and flowers can play a major role in export of these commodities.

xii. Employment opportunities have increased in marketing,transport, cold storage and warehousing facilities, credit, insuranceand logistic support services because of enhanced agriculturalproduction.

Successful Entrepreneurs of Jammu and Agri-Start-Ups Incubated at Agri-Business Incubator, SKUAST-Jammu

Various success stories of agri-entrepreneurs have been reportedfrom different sectors and states suggesting their growing importance inagricultural extension (Srinivas IAS et al., 2014). Below aresome handpicked success stories about the successful entrepreneurs, farm innovators and agri start-ups incubated at SKUAST-Jammu.

Successful Entrepreneurs / Farm Innovators

Mr. Kulbushan Khajuria (58) is a resident of Sohanjana village, Tehsil Mandal, District Jammu of Union Territory of Jammu and Kashmir. He owns 60 healthy milch animals and an automated milk collection system, a well-maintained cattle shed and a feeding area. Currently, his dairy named Krishna Dairy Farm has 48 Holstein Friesians (HF) Cows, nine Jersey cows, two Gir cows, one Shahiwal cow and one Murrah Buffalo. At present, 50 percent of his dairy animals are in the lactating phase and producing 4.5 quintals of milk every day. For marketing of milk, he has established 3 sale counters in Jammu city at TalabTillo, Shakti Nagar and ChanniHimmit Jammu. He is generating a turnover of more than 60 lakhs per year from sale of milk, vermin compost and agricultural products with a net profit of Rs 20-25 lakhs per year. He is providing employment to 7 unemployed youths round the year and 10 persons are engaged for need based services as and when required for cutting of fodder. He is also producing and selling 400-500 qtl.of vermin compost every year through sale counters. To minimize cost of milk production, he is preparing silage on his own farm every year. His visionary and leadership qualities have made him a successful entrepreneur. The success of Mr. Khajuria can be gauged from repayment of loan of 30 lakhs in 5 years that he had taken from the bank at the time of expansion of his dairy farm. He has a clear vision of his venture that helped him to turn the idea of establishing a dairy farm into reality by pooling of his personal resources and availing bank loan.

Sh. KuldeepVerma S/o Sh. Puran Chand is aresidentofvillageKotliCharakan, Jammu. He has constructed a pond of 1000 sq. metre. In the initial years, the fish production was less ranging between 200-300 kg/year which was encouraging but not upto his satisfaction. Then, he contacted scientists of KVK

SKUAST Jammu for technical details and scientific fish farming techniques. KVK Jammu laid front line demonstration of fisheries at his field and briefed him with all the latest technologies and calendar of activities to follow. He followed all the instructions and achieved fish production of 400kg during the year 2020. He is earning a net profit of Rs. 59000 per year from sale of fishes. Other than this, he is rearing backyard poultry birds near the fish pond after getting advised from KVK Jammu regarding integrated farming system and excreta of birds will work as feed for fishes and it will reduce his production cost. He is having Sahiwal cow for milk and two dogs for watch during night. His activities are continuously captured and recorded to motivate other youths.

Sh. Dunda Singh a turmeric farmer from Sultanpur village in Kathua district has three roles by profession- a farmer, an entrepreneur and a turmeric Cultivator. For setting up of processing plant he purchased land (1 kanal 3 marlas) @ 1.5 lakhs per kanal with the initial investment of Rs. 7.00 lakhs for creation of infrastructure, 6.92 lakhs on machinery and established the Baba Deep Singh Spices Park with Brand Name BDS at village Kana Chak and registered his products with Food Safety Standard Authority of India (FSSAI) vide Registration No 11017210000050.There is a handsome increase in his income after setting up of the processing plant. By putting all these stepping stones together he started his own turmeric processing plant in 2014 and named it as Baba Deep Singh Spices Park and BDS Haldi is one of the most trusted brands of his processing plant. His family including wife, two sons, two Daughters in laws and maternal uncle plays a major role in all the activities related to the processing of turmeric such washing, boiling, polishing and grinding. He also has seasonal laborers working for him in the processing plant and all the packaging, sealing and grinding work is done by the equal contribution of family members in his factory and home. All these steps taken by him has increased the production of raw turmeric from 60 Quintals to 110 Quintals today.

Startups incubated at Agri-Business Incubator, SKUAST-Jammu

Bhavishya Devgun

BhavishyaDevgun successfully running his start up 'SalutemRidegear' selling AGRITREAT – An organic agri-input fluid spray offered as an alternative to harmful chemical pesticides & fungicides after conducting successful trials with farmers pan India. He has current sales generation of around 3 Lakhs pan India with distribution partners across 8 states & counting.

Nikky Kumar Jha

Nikky Kumar Jha successfully established his start up 'Saptkrishi Scientific Pvt. Ltd. selling SABJI KOTHI – A solar powered, cost effective, IOT enabled, microclimate based wheel mountable Preservator for storage of fresh fruits and vegetables to extend their shelf life by 5-30 days for vendors and small producers. He has exhibited his innovation throughout the country and has been invited by Govt. of Arunachal Pradesh to run a pilot trial there.

Payal Sharma

Payal Sharma successfully running her start up 'Jk Aromatics & Herbs LLP' selling chemical free, natural, sustainable products such as incense, rapid compost, plantablerakhis and chemical free soaps made from flower waste collected from temples/mosques across J&K thereby converting waste to wealth. She has already established a loyal customer base of more than 500 customers.

Vivek Sharma

Vivek Sharma successfully running his start up 'Vatika Agro Services' selling high quality, high yield, disease free seeds & seedlings of vegetables & cash crops developed in soil less media to farmers as well as individual customers including off season seedlings raised in protected structures. He is rapidly establishing a loyal customer base in the Jammu province & is looking to soon expand his operations outside as well.

Kuldeep Dhaliwal

Kuldeep Dhaliwal a progressive farmer successfully running his start up 'KdhariwalAgrotech& IT Services Pvt. Ltd. Selling CROP RECHARGER – A natural Agri input liquid that enriches the soil with nutrients and increases the nutrient absorption capacity of plants as an alternative to chemical fertilizers. He has won numerous awards & certifications from various govt. & non govt. Organizations for his innovative invention and has a huge customer base of fellow farmers.

Policies

The task of promoting entrepreneurship was earlier given to different departments and government agencies. In 2014, the Prime Minister decided to dedicate an entire ministry i.e., Ministry of Skill Development and Entrepreneurship to build this sector as he felt that skill development required greater push from the government's side. Furthermore, the idea is to reach 500 million people by the year 2022 through gap-funding and skill development initiatives.

India is gradually building a robust startup ecosystem. In order to promote and support entrepreneurs, the government has created a Ministry (department) dedicated to helping new businesses. The important schemes wherein budding entrepreneurs can get funding or grant-in-aid for establishment and growth of their enterprise are explained below:

RKVY-RAFTAAR

- The Rashtriya Krishi VikasYojana (RKVY) is an important scheme of the Government of India, Ministry of Agriculture and Farmers' Welfare (MoA&FW), aimed at strengthening infrastructure in agriculture and allied areas. In order to promote agripreneurship and agribusiness by providing financial support and nurturing the incubation ecosystem, a new component under the revamped scheme RKVY-RAFTAAR has been launched in 2018-19. Sher-e-Kashmir University of Agricultural Sciences and Technology of Jammu got selected as one of the Incubation Centre under RKVY-RAFTAAR project and Agri Business Incubator, SKUAST-Jammu has been established in March 2019.
- It is an initiative to encourage young minds to come with innovative ideas in the field of agribusiness and to develop an agripreneurship ecosystem.
- Under this project there are two Agriprenurship Development Programmes namely UDGAM (Pre-Seed Stage Funding) and PRAGATI (Seed Stage Funding) with grant-in-aid to the tune of 5 lakhs and 25 lakhs, respectively.

Atal Innovation Mission (AIM)

It is Government of India's flagship initiative to create and promote a culture of innovation and entrepreneurship across the length and breadth of our country. It was setup in 2016.AIM's initiatives have played an important contributory role in the advancement of India from a position of 81 in the Global Innovation Index in 2015 to a position of 48 in 2020.AIM has adopted a holistic approach encompassing schools, universities, research institutions, industry, MSME, NGOs, Ministries, at district, state and national levels.

Major Initiatives by Atal Innovation Mission (AIM)

i. Atal Incubation Centres-Fostering world class startups in universities, institutions, private sector adding new dimensions of outcomes based scale up and monitoring to existing incubator model.

ii. Atal New India Challenges-Fostering product and service innovations in the country with national socio economic impact and aligning them to the needs of various sectors/ministries/industry.

iii. Atal Community Innovation Centre- To stimulate community centric innovation and ideas in the unserved/underserved regions of the country including Tier 2 and Tier 3 cities and rural India

iv. Building strategic Innovation partnerships with Public sector, Private Sector, Multinationals, and Global Country to Country Partnerships which include USA, UK, Germany, Russia, Singapore, Sweden, Israel, Denmark, Australia to name a few Atal Innovation Mission initiatives are key to further promoting the culture of innovation and giving expression to the innovative and entrepreneurial ideas of startups to enable development of one of the most Innovative Nations of the world.

NIDHI

National Initiative for Developing and Harnessing Innovations (NIDHI) is an umbrella program conceived and developed by the DST for nurturing ideas and innovations (knowledge-based and technology-driven) into successful startups.

NIDHI has following program components

i. Technology Business Incubator (TBI), Seed Support System (SSP), Accelerator, Centre of Excellence (CoE), Promoting and Accelerating Young and Aspiring Innovators and Startups (PRAYAS), Entrepreneur in Residence (EIR),

ii. The NIDHI-EIR programme provides Rs. 30,000 per month as fellowship to an aspiring or budding entrepreneur for pursuing a promising technology business idea.

iii. While Promoting and Accelerating Young and Aspiring technology entrepreneurs (PRAYAS), support upto Rs. 10 lakhs to enable translation of an innovative idea to a prototype is provided.

iv. Inclusive TBI is a three years duration initiative supported by the Department of Science Technology DST for educational institutions who are likely to foster innovation and entrepreneurship culture among the students, faculties, entrepreneurs, and nearby communities.

Startup India Initiative

The Prime Minister of India launched the Startup India Initiative in the year 2016. The idea is to increase wealth and employability by giving wings to entrepreneurial spirits. The government gives tax benefits to startups under this scheme and 798 applicants have made use of this scheme to date. The Department of Industrial Policy and Promotion is maintaining this initiative and is treating it as a long term project. Moreover, the overall age limit for startups has been increased from two years to seven years and for biotechnology firms, the age limit is ten years from the date of incorporation. It is one of the best government-sponsored startup schemes for entrepreneurs as it provides several concessions.

Startup India Seed Fund

On 16 January 2021, Prime Minister Sh. NarendraModi announced the launch of the 'Startup India Seed Fund' — worth INR 1,000 crores — to help startups and support ideas from aspiring entrepreneurs. Prime Minister said that the government is taking important measures to ensure that startups in India do not face any capital shortage.Financial assistance to startups is also the part of the initiative for proof of concept, prototype development, product trials, market entry, and commercialization. Under the initiative, the Seed Fund is disbursed to eligible startups through eligible incubators across.

ASPIRE: Aspire has been launched by the Indian government to set up a network of technology centres, incubation centres and to promote startups for innovation and entrepreneurship in the rural and agriculture-based industry. (https://msme.gov.in/sites/default/files/AS PIRE_EN.pdf)

New GenIEDC: The NSTEDB startup scheme by the Indian government will provide a limited, one-time, nonrecurring financial assistance, up to a maximum of INR 25 Lakhs. (http://www.newgeniedc-edii.in/)

SFAC: The quantum of SFAC Venture Capital Assistance will depend on the project cost and will be the lowest of the following: >26 per cent of the promoter's equity and >INR 50 Lakhs. (http://sfacindia.com/Procedure_For_VCA_S cheme.aspx)

AIM: AIM will provide a grant-in-aid of INR 10 Cr to each Atal Incubation Centre for a maximum of five years to cover the capital and operational expenditure cost in running the centre. (http://niti.gov.in/content/atal-incubation-centresaics#Source: Inc42, NASSCOM).

Some of the recent major Government initiatives in the sector are as follows:

- In October 2021, the Union Minister of Home Affairs and Corporation launched the 'Dairy Sahakar' scheme in Anand, Gujarat.

- Ministry of Civil Aviation launched the Krishi UDAN 2.0 scheme in October 2021. The scheme proposes assistance and incentive for movement of agri-produce by air transport. The Krishi UDAN 2.0 will be implemented at 53 airports across the country, largely focusing on Northeast and tribal regions, and is expected to benefit farmers, freight forwarders and airlines.
- In October 2021, Agricultural and Processed Food Products Export Development Authority (APEDA) signed a Memorandum of Understanding (MoU) with ICAR-Central Citrus Research Institute (ICAR-CCRI), Nagpur, for boosting exports of citrus and its value-added products.
- In September 2021, Prime Minister Mr. NarendraModi launched 35 crop varieties with special traits such as climate resilience and higher nutrient content.
- Prime Minister of India launched the Pradhan Mantri Kisan Samman Nidhi Yojana (PM-Kisan) and as per the Union Budget 2021-22, Rs. 65,000 crores (US$ 8.9 billion) was allocated to PradhanMantriKisanSammanNidhi (PM-Kisan).
- The Indian government has initiated Digital Agriculture Mission for 2021-25 for agriculture projects based on new technologies such as artificial intelligence, block chain, remote sensing and GIS technology, drones, robots and others.
- In September 2021, the Union Ministry of Agriculture and Farmers' Welfare signed five MoUs with CISCO, Ninjacart, Jio Platforms Limited, ITC Limited and NCDEX e-Markets Limited. This MoU will have five pilot projects, which will help farmers make decisions on the kind of crops to grow, variety of seeds to use and best practices to adopt to maximise yield.
- With a budget of US$ 1.46 billion, the 'Production-Linked Incentive Scheme for Food Processing Industry (PLISFPI)' has been approved to develop global food manufacturing champions commensurate with India's natural resource endowment and to support Indian food brands in international markets.
- Rs. 4,000 crores (US$ 551.08 million) was allocated in Union Budget 2021-22 towards implementing PradhanMantri Krishi SinchayeeYojana (PMKSY-PDMC).
- The Ministry of Food Processing has been allocated Rs. 1,308.66 crore (US$ 180.26 million) in the Union Budget 2021-22.

- Under PradhanMantriFormalisation of Micro Food Processing Enterprises (PM FME), an outlay of Rs. 10,000 crore (US$ 1.34 billion) over a period of five years from FY21 to FY25 has been sanctioned.
- To boost farmer incomes and growth of the agricultural economy, the Indian government released funds in June 2021 for farm mechanisation such as establishment of custom hiring centres, farm machinery bank and high-tech hubs in different states.
- In April 2021, the Government of India approved a PLI scheme for the food processing sector with an incentive outlay of Rs 10,900 crore (US$ 1,484 million) over a period of six years starting from FY22.
- The Government of India came out with Transport and Marketing Assistance (TMA) scheme to provide financial assistance for transport and marketing of agriculture products in order to boost agriculture exports.
- The Agriculture Export Policy, 2018 was approved by the Government of India in December 2018. The new policy aimed to increase India's agricultural export to US$ 60 billion by 2022 and US$ 100 billion in the next few years with a stable trade policy regime.
- The Government of India is going to provide Rs. 2,000 crores (US$ 306.29 million) for computerization of Primary Agricultural Credit Society (PACS) to ensure cooperatives are benefitted through digital technology.
- The Government of India launched the PradhanMantri Krishi SinchaiYojana (PMKSY) with an investment of Rs. 50,000 crores (US$ 7.7 billion) aimed at development of irrigation sources for providing a permanent solution from drought.
- Government plans to triple the capacity of food processing sector in India from the current 10% of agriculture produce and has also committed Rs. 6,000 crores (US$ 793 million) as investments for mega food parks in the country, as a part of the Scheme for Agro-Marine Processing and Development of Agro-Processing Clusters (SAMPADA).
- The Government of India has allowed 100% FDI in marketing of food products and in food product E-commerce under the automatic route.

Atmanirbhar Bharat Package for Farmers and Agri-entrepreneurs

S. No.	Particulars of Package	Amount (Rs. Cr.)
	The third tranche of Atmanirbhar Bharat Abhiyan	1,50,000
1.	Agri Infrastructure Fund for farm gate infrastructure including cold chain and post-harvest infrastructure	1,00,000
2.	Food micro-enterprises with a cluster-based approach	10,000
3.	PradhanMantriMatsyaSampadaYojana (PMMSY)	20,000
a)	For activities in Marine, Inland fisheries and Aquaculture	11000
b)	For Infrastructure - Fishing Harbours, Cold chain, Markets, etc.	9000
4	Animal Husbandry Infrastructure Development Fund	15,000
5	Promotion of Herbal Cultivation	4000
6	Extending Operation Greens project from Tomatoes, Onion, and Potatoes (TOP) to all fruits and vegetables	500
7	Beekeeping initiatives	500

Conclusion

Agri-entrepreneurship are facing various challenges in the execution of their plans to develop new start-ups or any other support businesses. The basic problem can be the literacy level of the agri-preneurs, farmers or rural people and another strong chance is lack of innovative mind. This acts as root cause for majority of challenges mentioned below. Major challenges faced by the agri-preneurs are inadequate infrastructural facilities, lack of innovative entrepreneurial abilities among people, declining interest in agricultural allied services, lack of skilled and managerial manpower, marketing problems, lack of awareness about career in agri-preneurship, lack of budget, lack of strategy, innovation initiatives with no innovation strategy, fear of making mistakes, cognitive barriers etc.

Agri-entrepreneurship is the need of the hour to make agriculture a more attractive and profitable business enterprise. Agriculture has great scope for entrepreneurship and innovative start-ups and it needs useful management of production and marketing parameters. Agriculture in developing countries is not successful to produce enough income to lift people out of their poverty and scarcity. Thus, innovation and agriculture definitely generates a solution for growing household incomes. The excellent management and entrepreneurial expertise infused with government measures would facilitate success of the growing needs of agri-businesses. Innovation, confidence, risk bearing, honesty and visionare the parameters to be adopted and follow for becoming a successful agri-entrepreneur. Agri-entrepreneurship provides direct employment especially in rural areas and contributes to the national income. Value addition of products offer entrepreneurs with high profit and return-on-investment which otherwise farmers is not able to receive. Therefore, farmers by becoming entrepreneurs have to take benefit of change in consumer demand and satisfy consumers' needs with value-added products.

References

Bansal A. 2011.Agri-Business in India–Vision 2020.International Journal of Multidisciplinary Research, 1(2):62-72.

Baregheh, A., Rowley, J. and Sambrook, S. 2009. Towards a multidisciplinary definition of innovation.Management Decision, 47(8):1323-1339.

Buttriss J. L. 2011. Feeding the planet: an unprecedented confluence of pressures anticipated Nutrition Bulletin 36, 235 – 241.

Dollinger, S. J. 2003. Need for uniqueness, need for cognition and creativity. The Journal of Creative Behavior, 37(2): 99–116.

Gray, C. 2002. Entrepreneurship, resistance to change and growth in small firms. Journal of Small Business and Enterprise Development,9(1):61-72.

Gupta, U. (2012). Customer Loyalty towards Kiranas in Competitive Environment; A Case study International Journal of Marketing and Technology, II (8), 249-268

Kashyap, P., Prusty, A.K., Panwar, A.S., Kumar, S., Punia, P., Ravisankar, N. Kumar, V. 2019. Agri-Entrepreneurship: Challenges and Opportunities, Edited Book published by Today and Tomorrow's Printers and Publishers, New Delhi.

Misra A., Singhal N., Sivakumar B., Bhagat N., Jaiswal A., Khurana L. 2011. Nutrition transition in India: secular trends in dietary intake and their relationship to diet-related non communicable diseases J. Diabetes 3, 278 – 292.

Popkin B. M. 2002. An overview on the nutrition transition and its health implications: the Bellagio meeting Pub. Health Nutr. 5, 93 – 103.

Popkin B. M., Adair L.S., Ng S. W. 2012. Global nutrition transition and the pandemic of obesity in developing countries Nutr. Reviews 7, 0 3 – 21.

Sah P, Sujan D K and Kashyap S K. 2009.Role of agripreneurship in the development of rural area, Paper presentation in ICARD at Banaras Hindu University, Varanasi.

Silayoi, P., and Speece, M. 2004. Packaging and purchase decisions: An exploratory study on the impact of involvement level and time pressure. British Food Journal, 106 (8), 607-628.

Srinivas, IAS., B. Shekara, P. C., Murthy, L. and Sahare J. 2014. Agri-Clinics and Agri-Business Centres Success Stories of Agripreneurs.National Institute of Agricultural Extension Management (An Organization of Ministry of Agriculture, Govt. of India). Rajendranagar, Hyderabad.

Sundar, I. 2016. Agribusiness Scope, Opportunities and Challenges in India. International Journal of Economic and Business Review. 4(7):171-178

7

Application of Artificial Intelligence in Animal Health

P. Krishnamoorthy[1] and K.P. Suresh[2]

[1]*Senior Scientist, Pathoepidemiology Laboratory, ICAR National Institute of Veterinary Epidemiology and Disease Informatics (NIVEDI), Ramagondanahalli, Yelahanka Bengaluru-560064, Karnataka*

[2]*Principal Scientist, Pathoepidemiology Laboratory, ICAR National Institute of Veterinary Epidemiology and Disease Informatics (NIVEDI), Ramagondanahalli, Yelahanka Bengaluru-560064, Karnataka*

Artificial intelligence is a computational science concerned with smart computer machines capable of performing tasks similar to the human intelligence. The term AI is widely used for the projects involved in the developing the intellectual processes of human intelligence such as reasoning ability, meaning discovery, generalization, learning from the past. AI technology mainly analyses the huge amount of data based on the set of instructions or commands known as algorithms to do a specific function or task. The mathematician Alan Turing with a simple question "Can machines think?" revolutionised the history. His paper "Computing Machinery and Intelligence" in 1950 and Turing test has favour the development of goal and the vision of AI.

The research in AI are mainly focussed on the following intelligence components namely :

a) Learning
b) Reasoning
c) Problem solving
d) Perception and
e) Language use

Learning involves different forms when applied to the AI. The simple method of learning by the computer is the trial and error method. The reasoning involves the depending on the situation how to draw the inferences or solutions or actions. The inferences may be divided in to deductive or inductive. Problem solving by AI is characterized by the systematic search through a wide range of possible actions to achieve a specified task or solution or goals. Examples are simple robot action such as pickup, putdown, move right, etc. The perception involves the scanning of the environment by means of sensory organs, which are real or artificial and the scene is divided in the different objects in various spatial relationships. Examples such as optical sensors to identify the individuals, robots roam to collect empty soda cans. The language used in AI is the system of signs having meaning by convention. Examples are, the traffic signs form a mini language, hazard ahead, those clouds mean rain.

AI addresses the three challenges in the animal health, a) understanding a situation and its dynamics, e.g., epidemic spread, b) the perception of the environment, which corresponds to the detection of patterns in animal health (e.g., repeated sequence of observations), forms (protein) and signals (increase mortality compared to a baseline) at different scales, c) computer-based decision making or more realistically, human decision support such as expert systems, diagnostic support, resource allocation (Ezanno et al., 2021).

There is also growing demand for the development and use of wearables, smart cameras and other sensors in animal health for pet animals and on farms. These devices used will generate more amount of data, which results in the potential for AI by using machine learning algorithms and real time data analysis. The data plays a key role in the development of the AI enabled solutions. The quality of data is more important rather than the quantity of data for the solutions derived by using AI and also emphasize the need to collect, organise and make accessible quality data at central point.

Applications of AI in Animal Health

The AI is widely used in the field of human health and gaining importance in the field of animal health also. It has wider applications in the animal health and some of them are given below.

1. AI in animal health helps the veterinary diagnostics easier, accessible medical care by all, and also provides the convenience in data collection.
2. Tool for disease diagnosis-it is used as a tool for disease diagnosis in animals and these programmes suggest the potential diagnosis, next step therapies or course of actions.
3. Imaging tool – organizes the images for the clinical interpretation, increases the efficiency significantly.
4. AI for the pet care and pet industries includes
    a) pet trackers and pet cameras,
    b) pet telehealth,
    c) dog walking,
    d) smart feeder and pet food,
    e) solutions for the pet world (Singh et al., 2021)

5. AI for livestock diseases prediction e.g., NADRES, BT Mobile app. AI and machine learning powered data capturing and forewarning system for livestock diseases are provided.
6. Diagnosing and managing Chronic kidney disease (CKD) in dogs and cats by analysing the available data and high accuracy of AI and machine learning in predicting the CKD.
7. It helps in the informed decision making based on the data rather than decision based on intuitions.
8. It identifies the patterns and forms of diseases in the animal population.
9. It is used in the development of expert systems for making realistic decisions and diagnosis of animal diseases.e.g., CaDDES.
10. AI driven digital platforms for the clinical disease diagnosis of livestock diseases using the image processing and matching for the prevention and early detection of livestock diseases.
11. Animal disease surveillance and digital veterinary services in India
12. Reporting of diseases and risk communication to field Veterinarians in the form of Short Messaging Service (SMS) or voice messaging.

Case studies

To understand better the application of AI in animal health, the below are the few case studies or works undertaken using AI in livestock diseases forewarning and diagnosis in India.

Case Study 1: National Animal Disease Referral Expert System (NADRES)

Early recognition of aserious or exotic animal disease can be done which is one of the most important factors influencing the chance of controlling the disease and reducing its economic and social impact on the whole country. In an attempt to look for a more reliable model, we developed adisease-climate relationship models using artificial intelligence and GIS to predict 13 economically important livestock disease outbreaks in India. Disease outbreaks data collected from 31 AICRP on ADMAS centres is stored and maintained in NADRES v2 database since year 1987. The risk factors data consisting of environmental variables such as precipitation, soil moisture, potential evaporation, wind speed, temperature, specific humidity, surface pressure was retrieved from GES DISC "GLDAS_NOAH025_M.2.1" dataset and remote sensing variables such as NDVI, LST (MODIS dataproduct). The livestock population data was obtained from 20th Livestock census 2019, Department of Animal Husbandry and Dairying (DAHD), New Delhi.The fitted models were assessed fortheir discriminating power using Receiving Operating Characteristic (ROC) curve, Cohen's Kappaand True SkillStatistics (TSS). The outcome of best fitted model/s were in probability of disease occurrence and was categorised into 6 risk levels-No risk (NR), Very lowrisk (VLR), Low risk (LR), Moderate risk (MR), High risk (HR)and Very high risk (VHR) are given in Figures 1 and 2. Theresults of forewarning is sent to allthe state animal husbandry departments and DAHD for taking necessary preventive measures in two months' advance to avoid the occurrence of disease incidence. Thefore warning results are generated at block level for 4 economically important diseases in Karnataka stateas pilot mode and was communicated to 2,000 veterinary officers by auto messaging to their mobiles. This web application provides several other features such asanalysis results of disease pattern ineach state, disease maps, distribution of livestock population, impact analysis etc. Epi calculator for developing sampling plan was incorporated in the application. All results in the application is dynamic and automated by using R software codes and it takes 2-3 days for results generation for a month (Suresh et al., 2019).

Fig. 1 NADRES Home page of the Website

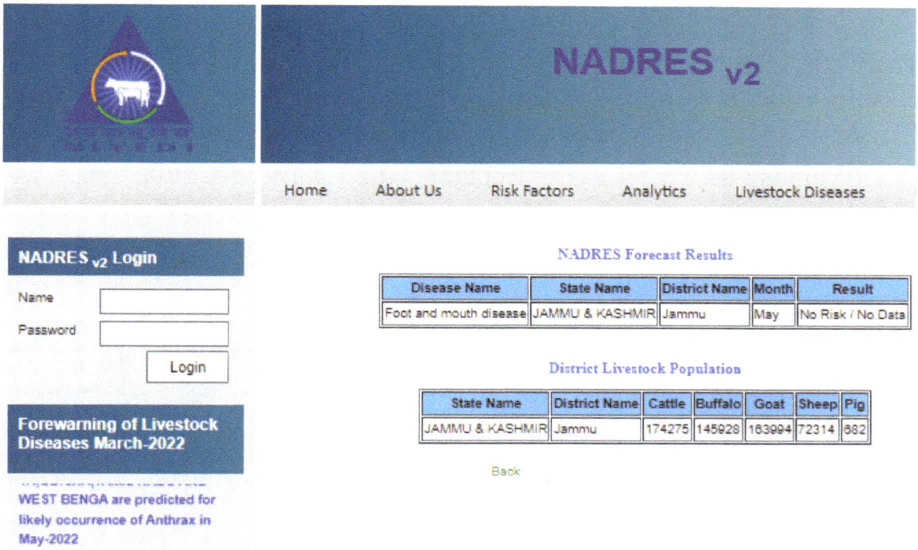

Fig. 2. Disease prediction page of NADRES v2

Case Study 2: Bluetongue Forewarning at Taluk Level for Karnataka Mobile App

The Bluetongue is a contagious, insect borne, viral disease affecting ruminants mainly sheep. In the NADRES, we are forecasting the diseases at district level for two months in advance. But in this mobile application, the disease forewarning is given at the taluk level for Karnataka state (Figure 3) and it helps in taking the preventive measures in less geographical area and includes the data obtained from past years' outbreaks, sero surveillance data, and more number of weather parameters. This was developed under the ICAR sponsored National Innovations in Climate Resilient Agriculture (NICRA) project funds.

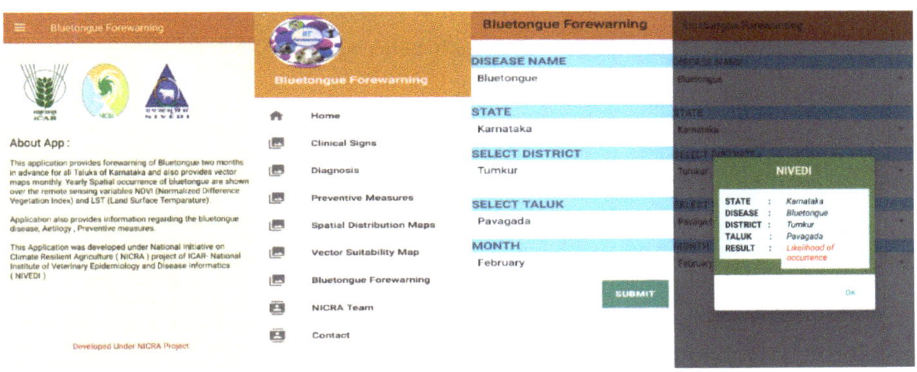

Fig. 3. Bluetongue forewarning mobile app screenshots

Case Study 3: Cattle Disease Diagnosis Expert System (CaDDES)

The CaDDES is a web application (Figure 4) developed for the diagnosis of thirteen cattle diseases in the field conditions using the twelve step process (Krishnamoorthy et al., 2019).

1. Identified clinical signs/symptoms for thirteen cattle diseases (Anaplasmosis, Anthrax, Babesiosis, Black quarter, Foot and Mouth Disease, Haemorrhagic septicaemia, Theileriosis, Trypanosomiasis, Rabies, Infectious Bovine Rhinotracheitis, Leptospirosis, Brucellosis and Mastitis).
2. Prepared questionnaire containing 52 signs/symptoms for the thirteen cattle diseases for collecting the scores based on Likert scale scoring (1-10).
3. Face validity of questionnaire by pilot survey conducted with veterinarians or experts.
4. Scoring (1-10) for clinical signs on the priority based importance in screening a cattle disease by Veterinarians.
5. Content validity was done by analysis of scores given by Veterinarians using Aiken's value.
6. Ranking of the scores for signs/symptoms.
7. Construction of weighted score matrix for each disease with weights assigned for each sign/symptom.
8. Prepared expert system by using the computer programming languages and algorithms by rule based methods.
9. Developed an expert system – Cattle Disease Diagnosis Expert System (CaDDES).
10. Internally Validated the web application.
11. External Validation at the field level for diagnosis of cattle diseases.
12. Development of mobile application for the benefit of Veterinarians.

It helps the veterinarian in the probable diagnosis of the diseases and further steps to be carried for confirmation of disease and to undertake preventive measures.

The AI has wider applications and there is need to explore the new avenues based on the innovations in ideas and thought process. The animals do have expressions, voice modulations, body language and memory for the experience as the humans and these factors aid in using the AI based solutions and software which can help in the early diseases diagnosis in animals and save the animals for better human health.

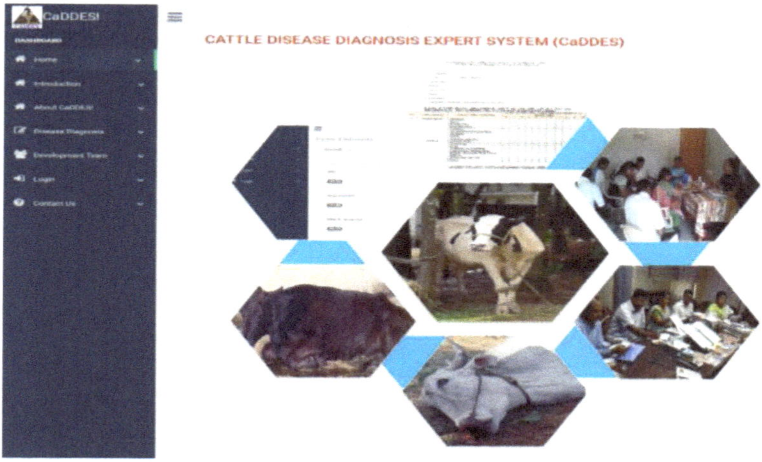

Fig. 4. Home of CaDDES web application

It also helps in the AI research and development on a long term basis in animal health. The highly changing scenario of the companies involved in animal health and AI sector whether international companies or start-ups, will able to provide better understanding and the interactions possible. It will also help in the development of linkages with the farmers, society and public or Government services. The opportunities for the possible collaborations with AI teams are required. There are few limitations and constraints of AI approaches such as training effort must be provided and generalized.The boom in the AI makes it possible in the integration of the knowledge and the many animal health players and welfare to upstream further.

References

Ezanno, P., Picault S, Beaunee, G, Bailly, X., Munoz, F., Duboz, R., Monod, H and Guegan, J.F. 2021. Research perspectives on animal health in the era of artificial intelligence. BMC Veterinary Research, 52: 40.

Krishnamoorthy, P., Suresh, K.P., Govindaraj, G and Roy P. Development of an expert system for cattle disease diagnosis: a participatory approach. ICAR NIVEDI Annual Report 2019. pp:9.

Singh, R. 2021. Application of artificial intelligence in animal health and veterinary services. PashudhanPraharee, Online August 23, 2021.

Suresh, K.P., Hemadri, D., Kurli, R., Dheeraj, R and Roy, P. Application of artificial intelligence for livestock disease prediction. Indian Farming, 69(3): 60-62.

8

Pet Practice Startup A Successful Model

Shilpi Das

Director, Companion Care Pet Clinic, Advisor, Cure and Care Pet Clinic New Delhi

India is the fastest-growing pet care market in the world and the pet population is of an estimate of 28-29 million, with dogs making up the bulk of pet ownership. The rising pet ownership has led to the growth of the pet care eco-system comprising retail chains, pet nutrition pharmaceuticals, grooming, toys, clothing, diagnostics, and clinical practice. The pandemic of Covid-19 has a significant impact on the Indian economy; and the pet industry has seen a steep growth during the pandemic period. Isolation and anxiety during the lockdown led to many restorations to adopt furry companions. The veterinary graduate has a lot of opportunities, and it depends upon the individual as to what sector the student wants to pursue. Employment opportunities include diverse areas such as clinical practice, teaching, and research, regulatory medicines, public health, wildlife specialist, or lab animals. Depending upon the interest, one can start exploring options to work for the pet industry and join the Veterinary pet practice clinical.

To start a pet practice business, one must wear two hats i.e. one as a good clinician and the other as a businessperson. It is important that one has to prepare themselves and develop an interest in clinical practice and be more comfortable with Veterinary medicine. Besides, developing clinical skills it is important to have marketing or business skills too. Once both Veterinary and business skills are developed, it's time to start some research and do some background work.

- Develop a business plan: Initial planning may take time and a good business plan will have a clear set direction for running the clinic. Identify the client base, identify staff, train your staff to deliver a quality customer experience.
- Competition Research: Before starting the practice, it is important to know about competition and to have knowledge of nearby clinics, the offer they are providing, and how different will be your clinic.

- Financial planning: It is imperative to have good financial planning. Understand what the cost would be to set up the clinic and operational cost. Assume, a no-profit period of 4 to 6 months and keep sufficient finance to run the clinic need.
- Legal formalities: Understand legal compliance obligations and necessary registration required. State Veterinary council registration is compulsory. If diagnostics like Ultrasound or X-Ray are placed necessary approvals will be required.
- Build a team: Build a dedicated team

A veterinary clinic can be a small clinic setup, polyclinic, or a hospital. A small clinic can be a small setup and is usually managed by a single doctor with a limited staff, a polyclinic is a larger setup with more diagnostic facilities while a hospital can be a large setup with multiple departments including inpatient dept and providing service 24x7. There may be different facilities a private clinic set up may have, like as below:

- OPD area: A clean, spacious OPD area to cater to multiple cases at a time
- Diagnostic facility: Diagnostics improve the effectiveness of treatments and avoid long-term complications. A Veterinary setup may have its diagnostics or laboratory facility or outsource it. Various laboratories collect the sample from clinics, and a tie-up with such laboratory will be helpful for the setup.
- The facility of Ultrasound and X-Ray: Having aUltrasound and X-Ray at the same setup is always good as it is convenient for the pet parent not to run from one clinic to another diagnostic center. It is, however, important to understand the certification that will be required for establishing these facilitiesand train oneself to use thisequipment.
- Other facilities may be a microchipping facility, dental scaling, ECG, pet store, and medicines, wellness center
- Grooming facility: Pet grooming industry is growing at a fast pace, and the pet parents want their pets to feel fresh and keep them tidy. Pet grooming is also necessary for early identification and treatment of skin infection. A proper grooming center will have a bathing tub with a good water facility, and a powerful dryer to dry up the pet hair.

The Pet industry is growing at a fast phase and as per India Pet Care Market Overview, 2021-2026 it is expected to grow at a value of INR 7500 crores by the end of 2025. Hence, Veterinarians have a huge scope to enter clinical practice and be an entrepreneur.

9

Entrepreneurship Wings in Veterinary Sector

Ankur Sharma

Assistant Professor, Division of Veterinary Surgery and Radiology
Faculty of Veterinary Science and Animal Husbandary, R.S. Pura
Sher-e-Kashmir University of Agricultural Sciences and Technology
Jammu, India

Entrepreneurship opportunities in veterinary sciences have expanded significantly with the integration of technology. The convergence of veterinary expertise and innovative technological solutions has opened up new avenues for entrepreneurs to create impactful businesses in the field. This field offers a wide range of opportunities for persons passionate about animal health and welfare. In the recent years, the demand for high-quality veterinary services and products has increased tremendously, creating an enriching environment for aspiring entrepreneurs. There are various avenues available in the veterinary industry for individuals thinking about establishing their own businesses.

Establishing a Veterinary Clinic or Hospital

One of the most evident entrepreneurial ventures in veterinary sciences is the establishment of a veterinary clinic or hospital. With an increasing number of pet owners seeking professional care for their animals, there is a constant demand for well-equipped and reputable veterinary facilities. Starting a veterinary clinic requires careful planning, including obtaining the necessary licenses, hiring skilled veterinarians and staff, and investing in diagnostic equipment and medical supplies. Additionally, offering specialized services, such as ophthalmology, dentistry, dermatology, or alternative therapies that can differentiate one's clinic from competitors.

Mobile Veterinary Services

In today's fast-paced world, convenience is key. Mobile veterinary services have gained popularity as they provide pet owners with the convenience of veterinary care right at their doorstep. Starting a mobile veterinary service allows entrepreneurs to reach a broader customer base and offer flexible appointment options. This business model requires a well-equipped vehicle and the ability to offer a range of services on the go. Providing vaccinations, check-ups, and basic treatments can help establish a good customer base and expand the business further.

Pet Rehabilitation and Physical Therapy

With the growing awareness of pet health and the need for specialized care, the field of pet rehabilitation and physical therapy has emerged as a promising entrepreneurial opportunity. This field focuses on assisting animals in recovering from surgeries, injuries, or chronic conditions through various physical therapy techniques. Entrepreneurs can establish rehabilitation centers equipped with state-of-the-art equipment, such as underwater treadmills and laser therapy machines, and hire certified therapists to provide rehabilitation services to pets.

Pet Product Manufacturing

The pet industry is thriving, and pet owners are increasingly seeking high-quality products for their beloved companions. From pet food and treats to grooming products and accessories, there is a vast market for innovative and sustainable pet products. Entrepreneurs can capitalize on this opportunity by starting their own pet product manufacturing business. By identifying gaps in the market and developing unique and appealing products, entrepreneurs can establish their brand and capture a significant market share.

Data Analytics and Artificial Intelligence (AI) in Animal Health

The vast amount of data generated in veterinary practices holds tremendous potential for entrepreneurial ventures. By leveraging data analytics and AI, entrepreneurs can develop innovative solutions that improve diagnosis, treatment, and overall animal health management. AI algorithms can analyze medical records, lab results, and imaging data to assist veterinarians in making accurate diagnoses and personalized treatment plans. Entrepreneurs can also create AI-powered wearable devices that monitor vital signs, behavior patterns, and disease indicators, providing proactive health management for pets.

Veterinary E-commerce Platforms

The e-commerce revolution has significantly impacted the pet industry, offering entrepreneurs an opportunity to establish veterinary-focused online marketplaces. These platforms can connect pet owners with a wide range of products, including pet food, medications, grooming supplies, and specialized veterinary equipment. By providing a convenient one-stop-shop experience, entrepreneurs can tap into the growing demand for online pet-related products and services. Building strong partnerships with suppliers and ensuring efficient logistics can be critical to the success of such ventures.

Mobile Apps for Pet Health Management

Entrepreneurs can develop mobile applications that empower pet owners to manage their pets' health effectively. These apps can provide features such as vaccination reminders, medication tracking, nutrition advice, and behavioral training programs. Additionally, integrating wearable device data into the apps can allow pet owners to monitor their pets' activity levels, sleep patterns, and overall well-being. Such apps not only enhance the bond between pet owners and their animals but also provide valuable insights for veterinarians to make informed decisions.

Veterinary Education and Training

Entrepreneurship opportunities extend beyond clinical services and products in veterinary sciences. Entrepreneurs can establish online platforms or e-learning programs that offer veterinary education and training to aspiring professionals. These platforms can provide courses, webinars, and interactive modules covering various topics such as veterinary medicine, surgery techniques, and animal behavior. By leveraging technology, entrepreneurs can democratize access to quality education and help bridge the gap between demand and supply of skilled veterinary professionals

Telemedicine and Teleconsultation

Advancements in technology have revolutionized veterinary industry. Telemedicine and teleconsultation services in veterinary sciences have gained traction, allowing pet owners to seek professional advice and guidance remotely. Entrepreneurs can establish online platforms that connect pet owners with licensed veterinarians, facilitating remote consultations, diagnosis, and treatment. These platforms can incorporate features such as video calls, chatbots, and digital medical records, enabling pet owners to access convenient and timely veterinary care from the comfort of their homes, e.g. wiggles, veteva, etc.

The field of veterinary sciences offers a myriad of exciting entrepreneurial opportunities for individuals passionate about animal health and welfare. From establishing veterinary clinics and mobile services to venturing into pet rehabilitation, data analytics, e-commerce, mobile apps, and educational initiatives, product manufacturing, and telemedicine, the possibilities are vast. Entrepreneurs in the veterinary industry must combine their passion for animals with strong business acumen to succeed. By recognizing the evolving needs of pet owners and providing high-quality services and products, entrepreneurs in this field have the potential to revolutionize the way veterinary services are delivered, improve the overall well-being of animals, and create successful businesses in the process, while making a positive impact on animal lives.

10

Finance and Funds for Entrepreneurship and Innovation

Anil Bhat and Malika Sharma

Assistant Professor, Division of Agricultural Economics and Agribusiness Management, Sher-e-Kashmir University of Agricultural Sciences and Technology, Jammu, Jammu and Kashmir, India

Entrepreneurship involves seeking and exploiting opportunities available for the benefit of the society. The changing agro-economic conditions in our country have offered a prospective context for developing entrepreneurship in agri-business in the country. Agri-entrepreneurship has the prospect of social and economic development, for example, employment generation, poverty reduction, improvements in nutrition, health and overall food security in the national economy especially in rural areas. The task of promoting entrepreneurship was earlier given to different departments and government agencies. In 2014, the Prime Minister decided to dedicate an entire ministry to build this sector as he felt that skill development required greater push from the government's side. The idea is to reach maximum people and startups through funding and skill development initiatives.

There are number of schemes which are currently financing for promotion of agri startups and agri-preneurship. Details of few among them are given below:

(i) **Seed Capital Fund Scheme**: Government of Jammu and Kashmir UT, Department of Labour and Employment (Source: JKEDI)

Seed Money

Apart from imparting skill based training to aspiring entrepreneurs from various economic sectors, the institute is also implementing a major finance support programme, Seed Capital Fund Scheme (SCFS), of the State government. SCFS is an important component of "Sher-e- Kashmir Employment and Welfare Programme for Youth (SKEWPY)" which in turn is a path breaking initiative of the government of Jammu and Kashmir for addressing unemployment in the state. Under SCFS, seed money of up to Rs 7.5 lakh is provided to eligible youth. The seed capital is like a onetime grant given to the entrepreneurs for making their business plans bankable. The rest of the business cost is provided through bank finance on low interest rates. JKEDI is having a tie up with the J&K Bank for hassle free loan for entrepreneurs promoted by them.

The main objective of the Scheme is to motivate, train and facilitate a large segment of educated young men and women to take up entrepreneurship as a career option instead of hankering after the government jobs. By opting for self employment a class of first generation entrepreneurs will be created who will create job opportunities not only for themselves, but also for others. It is also an objective of the Scheme to invest in various areas of economy to optimally exploit the resources.

Implementing Agency

The Department of Labour and Employment shall be the Administrative Department of J&K Entrepreneurship Development Institute for the purpose of Seed Capital Fund Scheme. The Scheme shall be implemented through Jammu & Kashmir Entrepreneurship Development Institute (JKEDI)

Seed Capital Fund

Seed Capital equivalent to 35% of the project cost shall be provided to the prospective entrepreneurs' upto a maximum of Rs. 4.00 lacs for under graduates, Rs. 5.00 lacs for graduates, Rs. 6.50 lacs for post graduates and Rs. 10.00 lacs for Professional/Technical Graduates. However, for Group initiatives the upper limit shall be relaxable upto sum total of individual entitlements. The beneficiaries for the scheme to have professional / technical qualification shall include the following categories of degrees:

i. Five and Four-Year Bachelors Degree Courses viz: MBBS, BDS, BVSc & A.H, BAMS, BUMS, BE, B.Tech, BSc (Agriculure), B.Sc (Forestry), LLB (including LLB done at University level after Graduation) etc. approved by concerned regulatory Body (s).
ii. Computer Science & Technology Graduates, MBA, etc. from recognized University / Institute.
iii. All two year Masters Degree Courses that are declared as professional degrees by Kashmir/Jammu Universities.
iv. For all such degrees that are awarded by other Universities and the holders whereof claim to possess professional degrees, shall be referred to respective Universities for clarification. However, in case of Professional/ Technical degree holders, the Seed Capital shall be available only if their proposed enterprise(s) is/are related to their qualifications. Otherwise they shall be treated at par with other graduates or post graduates as the case may be.

Eligibility condition for the scheme is as under:

a) A State subject in the age group of 18 to 40 years having a qualification of 10+2 or above who is unemployed,

b) There will be no income ceiling for coverage under the scheme;

c) The beneficiaries must be registered with the concerned District Employment and Counselling Centre. They can also be registered with other departments and institutions like DICs etc. wherever required so as to take advantage of the benefits which may accrue through such registrations. The JKEDI shall facilitate this during the course of training under Entrepreneurship Development Programme;

d) Any person or any, existing units and the units which have already availed of any incentive or subsidy under any scheme of the Government of India or the State Government shall not be eligible for assistance under the scheme;

e) No institution, corporate body, society or an NGO shall be eligible for assistance under the scheme.

Provided that for differently/specially abled persons, the age group of 18 to 42 years shall be applicable. Provided that the "person with disability" shall mean a person suffering from not less than forty percent of any disability, as certified by a competent medical authority.

(ii) Youth Startup Loan Scheme, Government of Jammu and Kashmir UT, Department of Labour and Employment

Start your business 0% hassle with Jammu & Kashmir Entrepreneurship Development Institute (JKEDI) and Mercy Corps under the overall supervision of Labour & Employment Department, Government of Jammu and Kashmir, UT. The institute also provides direct off-bank finance to eligible youth under its Youth Start up Loan Scheme (YSLS) programme whereby projects costing up to Rs.8 lakh are directly financed at 6% simple interest. Entrepreneurs covered under either of the programmes have to undergo mandatory three-week training. The training pro-gramme has been formulated with the help of international organizations like Making Cents whereby trainees are trained through experiential training methods and simulation exercises. With most of the process same, it differs with SCFS only on the resource front. Instead of free money, YSLS offers a loan to a maximum of Rs 8 lakh for start ups at only six percent of simple interest and that can cover up to 90% of the project cost.

The eligibility criteria under the Scheme are same as applicable under Seed Capital Fund Scheme which includes:

i. A State Subject in the age group of 18-37 years having a qualification of 10+2 or above.
ii. The beneficiary must be unemployed and registered with the concerned District Employment and Counseling Center.
iii. There shall be no income ceiling for coverage under the Scheme.
iv. Only new entrants to the field of entrepreneurship will be covered under the Scheme. Any person or any unit existing and the units which have already availed of any incentive or subsidy under any Scheme of the Government of India or the State Government shall not be eligible for assistance under the Scheme.
v. No Institution, Corporate Body, Society or NGO shall be eligible under the Scheme.

Sanctioning of Loan

The loan amount will be sanctioned in favour of the beneficiaries by the competent authority as per the benchmarks detailed out in the Business Plan. However, in case of projects with an investment of more than Rs. 8.00 lacs, the concerned beneficiaries shall have the obligation of investing their own contribution in the first phase. Thereafter the loan amount will be released in favour of the concerned beneficiaries.

Mode of Repayment

The total loan amount along with the interest @ 6% per annum (Simple Rate of Interest) shall be repayable within a period of 5 years in 20 equal quarterly installments. The repayment of loan shall have moratorium period of 6 months. During the moratorium period only interest charged shall be recovered on quarterly basis and the repayment of the principal with interest shall be taken up thereafter. The repayment received will be appropriated first towards interest and thereafter toward the principal. For this purpose, the beneficiary will be required to submit 20 post dated Account Payees Cheques from Jammu & Kashmir Bank Limited in his service area along with the Bank verification letter and attestation of signatures as per the prescribed format. The beneficiary shall have to produce original documents along with the attested photocopies of all the relevant documents required for record and verification.

Security

Since the scheme is being implemented on pilot basis, implementing agency shall seek a security from the beneficiaries against the Loan to be disbursed. The norms of security shall be such that the beneficiaries are able to meet them and the recoveries are ensured. In general, following security provisions shall be followed:

a) For Loans upto Rs. 4.00 lacs: Guarantee of a State Government Employee or Bank Employee with a service of at least 6 years at his/her disposal.

b) For Loans above Rs. 4.00 lacs: Guarantee of two State Government employees or Bank with a service of at least 6 years at his/her disposal.

Other Terms and Conditions

The beneficiary shall submit the following documents on the prescribed format duly notarized:

a) D.P Note

b) Deed of Irrevocable Power of Attorney

c) Deed of Guarantee (the guarantor/s should be a Government Servant/Bank Employee) with at least 6 years of service to superannuate.

d) Deed of Hypothecation of assets created in the joint names of Jammu and Kashmir Entrepreneurship Development Institute and the State Government.

e) Deed of Agreement

f) Affidavit

g) Letter of Undertaking.

(iii) RKVY-RAFTAAR

The Rashtriya Krishi Vikas Yojana (RKVY) is an important scheme of the Government of India, Ministry of Agriculture and Farmers' Welfare (MoA&FW), aimed at strengthening infrastructure in agriculture and allied areas. In order to promote agripreneurship and agribusiness by providing financial support and nurturing the incubation ecosystem, a new component under the revamped scheme RKVY-RAFTAAR has been launched in 2018-19. Sher-e-Kashmir University of Agricultural Sciences and Technology of Jammu got selected as one of the Incubation Centre under RKVY-RAFTAAR project and Agri Business Incubator, SKUAST-Jammu has been established in March 2019.

- It is an initiative to encourage young minds to come with innovative ideas in the field of agribusiness and to develop an agripreneurship ecosystem.
- Under this project there are two Agriprenurship Development Programmes namely UDGAM (Pre-Seed Stage Funding) and PRAGATI (Seed Stage Funding) with grants-in-aid to the tune of 5 lakhs and 25 lakhs, respectively.

(iv) Atal Innovation Mission (AIM)

Atal Innovation Mission is Government of India's flagship initiative to create and promote a culture of innovation and entrepreneurship across the length and breadth of our country. It was setup in 2016. AIM's initiatives have played an important contributory role in the advancement of India from a position of 81 in the Global Innovation Index in 2015 to a position of 48 in 2020.AIM has adopted a holistic approach encompassing schools, universities, research institutions, industry, MSME, NGOs, Ministries, at district, state and national levels.

Major Initiatives by Atal Innovation Mission (AIM)

Atal Incubation Centres-Fostering world class startups in universities, instituions, private sector adding new dimensions of outcomes based scale up and monitoring to existing incubator model.

1. Atal New India Challenges-Fostering product and service innovations in the country with national socio economic impact and aligning them to the needs of various sectors/ministries/industry.
2. Atal Community Innovation Centre- To stimulate community centric innovation and ideas in the unserved /underserved regions of the country including Tier 2 and Tier 3 cities and rural India
3. Building strategic Innovation partnerships with Public sector, Private Sector, Multinationals, and Global Country to Country Partnerships which include USA, UK, Germany, Russia, Singapore, Sweden, Israel, Denmark, Australia to name a few
4. Atal Innovation Mission initiatives are key to further promoting the culture of innovation and giving expression to the innovative and entrepreneurial ideas of startups to enable development of one of the most Innovative Nations of the world.

(v) NIDHI

National Initiative for Developing and Harnessing Innovations (NIDHI) is an umbrella program conceived and developed by the DST for nurturing ideas and innovations (knowledge-based and technology-driven) into successful startups.

NIDHI has Following Program Components

Technology Business Incubator (TBI), Seed Support System (SSP), Accelerator, Centre of Excellence (CoE), Promoting and Accelerating Young and Aspiring Innovators and Startups (PRAYAS), Entrepreneur in Residence (EIR).

The NIDHI-EIR programme provides Rs. 30,000 per month as fellowship to an aspiring or budding entrepreneur for pursuing a promising technology business idea.

While Promoting and Accelerating Young and Aspiring technology entrepreneurs (PRAYAS) will support upto Rs. 10 lakhs to enable translation of an innovative idea to a prototype.

Inclusive TBI is a three years duration initiative supported by the Department of Science Technology DST for educational institutions who are likely to foster innovation and entrepreneurship culture among the students, faculties, entrepreneurs, and nearby communities.

(vi) Startup India Initiative

Startup India is a flagship initiative of the Government of India, intended to catalyse startup culture and build a strong and inclusive ecosystem for innovation and entrepreneurship in India. The START UP INDIA initiative was launched in 2016 by the Honorable Prime Minister with the aim of supporting a thriving ecosystem of entrepreneurs and startups. The most important feature of START UP India is that it is a successful initiative in changing people's minds, creating an India of job creators rather than job seekers. It is this mindset that lifts the spirits of young people because it gives them confidence that if the Prime Minister of a country shows faith in the youth, it is the responsibility of the young people to stand up and demonstrate that belief exactly. If we glance at the statistical data, the number of startups in India has grown to more than 60,000 from less than 500 just 5 years ago, across 55 industry sectors. In 2021, India granted 28,000 patents while 2,500,000 trademarks were registered in 2020-21 compared to 70,000 in 2013. In July 2021, the department for promotion of industry and internal trade (DPIIT) had recognised 52,732 Indian startups.

Even in COVID scenario when many people lose their job, these start up remain as one of their hope and Government do played its part even in that case by promoting VOCAL FOR LOCAL. This provided the push so required by the youth, the moral boost to take up any job that they are capable to perform as they have this mind setup that Startup Initiative is there with them to hold the hands.

(vii) Startup India Seed Fund

On 16 January 2021, Prime Minister Narendra Modi announced the launch of the 'Startup India Seed Fund'— worth INR 1,000 crores — to help startups and support ideas from aspiring entrepreneurs. PM Modi said that the government is taking important measures to ensure that startups in India do not face any capital shortage. Under this scheme, financial assistance is provided to startups for proof of concept, prototype development, product trials, market entry, and commercialization. The Seed Fund will be disbursed to eligible startups through eligible incubators across India

References

Bairwa, Shoji Lal & Lakra, Kerobim & Kushwaha, Saket & Meena, Lokesh & Kumar, Pravin. 2014. Agripreneurship Development as a Tool to Upliftment of Agriculture. International Journal of Scientific and Research Publication. 4.

InfoDev. 2013. Agribusiness Incubation: Trainers Manual (available at https://www.infodev.org/infodev-files/ m12_traineemanual.pdf).

Kashyap, P., Prusty, A.K., Panwar, A.S., Kumar, S., Punia, P., Ravisankar, N. Kumar, V. 2019. Agri-Entrepreneurship: Challenges and Opportunities, Today and Tomorrow's Printers and Publishers, New Delhi.

Kaur, M., Bhat, A., Sadik-Zada, E.R. and Sharma, R. 2022. Productivity Analysis and Employment Effects of Marigold Cultivation in Jammu, India. Horticulturae, 8,263.

FSC
www.fsc.org
MIX
Papier | Fördert
gute Waldnutzung
FSC® C083411

Zeitfracht Medien GmbH
Ferdinand-Jühlke-Straße 7
99095 Erfurt, Deutschland
produktsicherheit@kolibri360.de